Quantitative Applications in the Social Sciences

A SAGE PUBLICATIONS SERIES

1. Analysis of Variance, 2nd Edition *Iversen/ Norpoth*
2. Operations Research Methods *Nagel/Neef*
3. Causal Modeling, 2nd Edition *Asher*
4. Tests of Significance *Henkel*
5. Cohort Analysis, 2nd Edition *Glenn*
6. Canonical Analysis and Factor Comparison *Levine*
7. Analysis of Nominal Data, 2nd Edition *Reynolds*
8. Analysis of Ordinal Data *Hildebrand/Laing/Rosenthal*
9. Time Series Analysis, 2nd Edition *Ostrom*
10. Ecological Inference *Langbein/Lichtman*
11. Multidimensional Scaling *Kruskal/Wish*
12. Analysis of Covariance *Wildt/Ahtola*
13. Introduction to Factor Analysis *Kim/Mueller*
14. Factor Analysis *Kim/Mueller*
15. Multiple Indicators *Sullivan/Feldman*
16. Exploratory Data Analysis *Hartwig/Dearing*
17. Reliability and Validity Assessment *Carmines/Zeller*
18. Analyzing Panel Data *Markus*
19. Discriminant Analysis *Klecka*
20. Log-Linear Models *Knoke/Burke*
21. Interrupted Time Series Analysis *McDowall/McCleary/Meidinger/Hay*
22. Applied Regression *Lewis-Beck*
23. Research Designs *Spector*
24. Unidimensional Scaling *McIver/Carmines*
25. Magnitude Scaling *Lodge*
26. Multiattribute Evaluation *Edwards/Newman*
27. Dynamic Modeling *Huckfeldt/Kohfeld/Likens*
28. Network Analysis *Knoke/Kuklinski*
29. Interpreting and Using Regression *Achen*
30. Test Item Bias *Osterlind*
31. Mobility Tables *Hout*
32. Measures of Association *Liebetrau*
33. Confirmatory Factor Analysis *Long*
34. Covariance Structure Models *Long*
35. Introduction to Survey Sampling *Kalton*
36. Achievement Testing *Bejar*
37. Nonrecursive Causal Models *Berry*
38. Matrix Algebra *Namboodiri*
39. Introduction to Applied Demography *Rives/Serow*
40. Microcomputer Methods for Social Scientists, 2nd Edition *Schrodt*
41. Game Theory *Zagare*
42. Using Published Data *Jacob*
43. Bayesian Statistical Inference *Iversen*
44. Cluster Analysis *Aldenderfer/Blashfield*
45. Linear Probability, Logit, and Probit Models *Aldrich/Nelson*
46. Event History Analysis *Allison*
47. Canonical Correlation Analysis *Thompson*
48. Models for Innovation Diffusion *Mahajan/ Peterson*
49. Basic Content Analysis, 2nd Edition *Weber*
50. Multiple Regression in Practice *Berry/ Feldman*
51. Stochastic Parameter Regression Models *Newbold/Bos*
52. Using Microcomputers in Research *Madron/Tate/Brookshire*
53. Secondary Analysis of Survey Data *Kiecolt/Nathan*
54. Multivariate Analysis of Variance *Bray/Maxwell*
55. The Logic of Causal Order *Davis*
56. Introduction to Linear Goal Programming *Ignizio*
57. Understanding Regression Analysis *Schroeder/Sjoquist/Stephan*
58. Randomized Response *Fox/Tracy*
59. Meta-Analysis *Wolf*
60. Linear Programming *Feiring*
61. Multiple Comparisons *Klockars/Sax*
62. Information Theory *Krippendorff*
63. Survey Questions *Converse/Presser*
64. Latent Class Analysis *McCutcheon*
65. Three-Way Scaling and Clustering *Arabie/Carroll/DeSarbo*
66. Q Methodology *McKeown/Thomas*
67. Analyzing Decision Making *Louviere*
68. Rasch Models for Measurement *Andrich*
69. Principal Components Analysis *Dunteman*
70. Pooled Time Series Analysis *Sayrs*
71. Analyzing Complex Survey Data, 2nd Edition *Lee/Forthofer*
72. Interaction Effects in Multiple Regression, 2nd Edition *Jaccard/Turrisi*
73. Understanding Significance Testing *Mohr*
74. Experimental Design and Analysis *Brown/ Melamed*
75. Metric Scaling *Weller/Romney*
76. Longitudinal Research, 2nd Edition *Menard*
77. Expert Systems *Benfer/Brent/Furbee*
78. Data Theory and Dimensional Analysis *Jacoby*
79. Regression Diagnostics *Fox*
80. Computer-Assisted Interviewing *Saris*
81. Contextual Analysis *Iversen*
82. Summated Rating Scale Construction *Spector*
83. Central Tendency and Variability *Weisberg*
84. ANOVA: Repeated Measures *Girden*
85. Processing Data *Bourque/Clark*
86. Logit Modeling *DeMaris*

Quantitative Applications in the Social Sciences

A SAGE PUBLICATIONS SERIES

87. **Analytic Mapping and Geographic Databases** *Garson/Biggs*
88. **Working With Archival Data** *Elder/Pavalko/Clipp*
89. **Multiple Comparison Procedures** *Toothaker*
90. **Nonparametric Statistics** *Gibbons*
91. **Nonparametric Measures of Association** *Gibbons*
92. **Understanding Regression Assumptions** *Berry*
93. **Regression With Dummy Variables** *Hardy*
94. **Loglinear Models With Latent Variables** *Hagenaars*
95. **Bootstrapping** *Mooney/Duval*
96. **Maximum Likelihood Estimation** *Eliason*
97. **Ordinal Log-Linear Models** *Ishii-Kuntz*
98. **Random Factors in ANOVA** *Jackson/Brashers*
99. **Univariate Tests for Time Series Models** *Cromwell/Labys/Terraza*
100. **Multivariate Tests for Time Series Models** *Cromwell/Hannan/Labys/Terraza*
101. **Interpreting Probability Models: Logit, Probit, and Other Generalized Linear Models** *Liao*
102. **Typologies and Taxonomies** *Bailey*
103. **Data Analysis: An Introduction** *Lewis-Beck*
104. **Multiple Attribute Decision Making** *Yoon/Hwang*
105. **Causal Analysis With Panel Data** *Finkel*
106. **Applied Logistic Regression Analysis, 2nd Edition** *Menard*
107. **Chaos and Catastrophe Theories** *Brown*
108. **Basic Math for Social Scientists: Concepts** *Hagle*
109. **Basic Math for Social Scientists: Problems and Solutions** *Hagle*
110. **Calculus** *Iversen*
111. **Regression Models: Censored, Sample Selected, or Truncated Data** *Breen*
112. **Tree Models of Similarity and Association** *James E. Corter*
113. **Computational Modeling** *Taber/Timpone*
114. **LISREL Approaches to Interaction Effects in Multiple Regression** *Jaccard/Wan*
115. **Analyzing Repeated Surveys** *Firebaugh*
116. **Monte Carlo Simulation** *Mooney*
117. **Statistical Graphics for Univariate and Bivariate Data** *Jacoby*
118. **Interaction Effects in Factorial Analysis of Variance** *Jaccard*
119. **Odds Ratios in the Analysis of Contingency Tables** *Rudas*
120. **Statistical Graphics for Visualizing Multivariate Data** *Jacoby*
121. **Applied Correspondence Analysis** *Clausen*
122. **Game Theory Topics** *Fink/Gates/Humes*
123. **Social Choice: Theory and Research** *Johnson*
124. **Neural Networks** *Abdi/Valentin/Edelman*
125. **Relating Statistics and Experimental Design: An Introduction** *Levin*
126. **Latent Class Scaling Analysis** *Dayton*
127. **Sorting Data: Collection and Analysis** *Coxon*
128. **Analyzing Documentary Accounts** *Hodson*
129. **Effect Size for ANOVA Designs** *Cortina/Nouri*
130. **Nonparametric Simple Regression: Smoothing Scatterplots** *Fox*
131. **Multiple and Generalized Nonparametric Regression** *Fox*
132. **Logistic Regression: A Primer** *Pampel*
133. **Translating Questionnaires and Other Research Instruments: Problems and Solutions** *Behling/Law*
134. **Generalized Linear Models: A United Approach** *Gill*
135. **Interaction Effects in Logistic Regression** *Jaccard*
136. **Missing Data** *Allison*
137. **Spline Regression Models** *Marsh/Cormier*
138. **Logit and Probit: Ordered and Multinomial Models** *Borooah*
139. **Correlation: Parametric and Nonparametric Measures** *Chen/Popovich*
140. **Confidence Intervals** *Smithson*
141. **Internet Data Collection** *Best/Krueger*
142. **Probability Theory** *Rudas*
143. **Multilevel Modeling** *Luke*
144. **Polytomous Item Response Theory Models** *Ostini/Nering*
145. **An Introduction to Generalized Linear Models** *Dunteman/Ho*
146. **Logistic Regression Models for Ordinal Response Variables** *O'Connell*
147. **Fuzzy Set Theory: Applications in the Social Sciences** *Smithson/Verkuilen*
148. **Multiple Time Series Models** *Brandt/Williams*
149. **Quantile Regression** *Hao/Naiman*
150. **Differential Equations: A Modeling Approach** *Brown*
151. **Graph Algebra: Mathematical Modeling With a Systems Approach** *Brown*
152. **Modern Methods for Robust Regression** *Andersen*
153. **Agent-Based Models** *Gilbert*
154. **Social Network Analysis, 2nd Edition** *Knoke/Yang*
155. **Spatial Regression Models** *Ward/Gleditsch*
156. **Mediation Analysis** *Iacobucci*
157. **Latent Growth Curve Modeling** *Preacher/Wichman/MacCallum/Briggs*
158. **Introduction to the Comparative Method With Boolean Algebra** *Caramani*
159. **A Mathematical Primer for Social Statistics** *Fox*
160. **Fixed Effects Regression Models** *Allison*

Series/Number 07-160

FIXED EFFECTS REGRESSION MODELS

Paul D. Allison

University of Pennsylvania

Los Angeles | London | New Delhi
Singapore | Washington DC

For information:

SAGE Publications, Inc.
2455 Teller Road
Thousand Oaks,
    California 91320
E-mail: order@sagepub.com

SAGE Publications India Pvt. Ltd.
B 1/I 1 Mohan Cooperative
    Industrial Area
Mathura Road, New Delhi 110 044
India

SAGE Publications Ltd.
1 Oliver's Yard
55 City Road
London EC1Y 1SP
United Kingdom

SAGE Publications Asia-Pacific Pte. Ltd.
33 Pekin Street #02-01
Far East Square
Singapore 048763

Printed in the United States of America

Library of Congress Cataloging-in-Publication Data

Allison, Paul David.
Fixed effects regression models / Paul D. Allison.
        p. cm. — (Quantitative applications in the social sciences; 160)
Includes bibliographical references and index.
ISBN 978-0-7619-2497-5 (pbk.)
    1. Regression analysis. 2. Social sciences—Statistical methods. I. Title.

HA31.3.A44 2009
519.5′36—dc22   2008044919

This book is printed on acid-free paper.

10   11   12   13   10   9   8   7   6   5   4   3   2

Acquisitions Editor:	Vicki Knight
Associate Editor:	Sean Connelly
Editorial Assistant:	Lauren Habib
Production Editor:	Brittany Bauhaus
Copy Editor:	QuADS Prepress (P) Ltd.
Typesetter:	C&M Digitals (P) Ltd.
Proofreader:	Victoria Reed-Castro
Indexer:	Marilyn Augst
Cover Designer:	Candice Harman
Marketing Manager:	Stephanie Adams

CONTENTS

About the Author   **vii**

Series Editor's Introduction   **ix**

1. Introduction   **1**

2. Linear Fixed Effects Models: Basics   **6**
    The Two-Period Case   7
    Extending the Difference Score Method
        for the Two-Period Case   10
    A First-Difference Method for Three or
        More Periods per Individual   12
    Dummy Variable Method for Two or More
        Periods per Individual   14
    Interactions With Time in the Fixed Effects Method   19
    Comparison With Random Effects Models   21
    A Hybrid Method   23
    Summary   26

3. Fixed Effects Logistic Models   **28**
    The Two-Period Case   28
    Three or More Periods   32
    Interactions With Time   37
    A Hybrid Method   39
    Methods for More Than Two Categories
        on the Response Variable   42
    Summary   47

4. Fixed Effects Models for Count Data   **49**
    Poisson Models for Count Data With
        Two Periods per Individual   49
    Poisson Models for Data With More
        Than Two Periods per Individual   54
    Fixed Effects Negative Binomial Models for Count Data   61
    A Hybrid Approach   65
    Summary   68

5. Fixed Effects Models for Events History Data   **70**
Cox Regression   71
Cox Regression With Fixed Effects   73
Some Caveats   77
The Hybrid Method for Cox Regression   79
Fixed Effects Event History Methods
    for Nonrepeated Events   79
Summary   85

6. Structural Equation Models With Fixed Effects   **87**
Random Effects as a Latent Variable Model   87
Fixed Effects as a Latent Variable Model   91
A Compromise Between Fixed Effects
    and Random Effects   92
Reciprocal Effects With Lagged Predictors   93
Summary   97

Appendix 1. Stata Programs for Examples in Chapters 2 to 5   **99**

Appendix 2. Mplus Programs for Examples in Chapter 6   **108**

References   **113**

Author Index   **116**

Subject Index   **118**

ABOUT THE AUTHOR

Paul D. Allison is Professor of Sociology at the University of Pennsylvania, where he teaches advanced graduate courses on event history analysis, categorical data analysis, and structural equation models with latent variables. He is the author of seven books and more than 50 journal articles. Every summer, he teaches 5-day workshops on survival analysis and logistic regression analysis that draw about 100 researchers from around the United States. A former Guggenheim Fellow, Allison received the 2001 Lazarsfeld Award for distinguished contributions to sociological methodology.

SERIES EDITOR'S INTRODUCTION

At a recent conference, I heard a presentation by a researcher analyzing country-year data with random effects models when fixed effects models should have been applied. The paper was well received by people from various social science backgrounds. There is clearly much confusion in many disciplines of the social sciences about how to choose between fixed and random effects models, and indeed, simply about what these models do. Allison's book is about arguably the more important and the more general of the two models, and it will fill a great need for a book in the Quantitative Application in the Social Sciences series on the topic, especially considering the greater availability of panel data these days.

The country-year example represents a form of data that have individual cases observed over time. Panel surveys are in vogue these days, for a good reason, because panel data allow the researcher to get at social dynamics, a must to truly understand social mechanisms. Whereas some panel surveys involve a yearly panel, like the British Household Panel Study Survey, which began in 1991 and is still going strong, others have just a few waves, such as the National Longitudinal Study of Adolescent Health in the United States, which conducted interviews in three waves between 1994 and 2002.

Whether the unit of analysis is a person, a firm, or a country, the errors in a regression model for each case will be correlated or dependent over time, usually because there exist unobserved characteristics that vary from one case to another. In such a situation, the assumption of independence of the errors for regression analysis is violated (here we restrict the discussion to the linear regression although the general principle also applies to regressions with limited dependent variables).

Both fixed effects and random effects models can solve the problem of correlated errors. But fixed effects models do much more. In Allison's words, these models "use each individual as his or her own control." By doing so, they actually *control* for all stable, unobserved variables, just as if these variables had been measured and included in the regression model. In that sense, these statistical models perform neatly the same function as random assignment in a designed experiment.

As we can see, models such as these are very useful for social scientists. It is our good fortune that my predecessor Michael Lewis-Beck supported Allison's proposal for a book in the series as the topic will be an essential extension to the linear regression and to regression models with limited or categorical dependent variables covered in many of the books in our series. The author has been a consistent contributor to social science methodology for the past three decades on many important topics, including two useful contributions in the series, *Missing Data* (2001) and *Event History Analysis* (1984), which still is a benchmark that other books introducing event history type of analysis in the social sciences are measured against. In the current volume, Allison introduces, in lucid language, various types of fixed effects models—those for continuous, categorical, count dependent variables, and in structural equation settings—as well as discusses the choice between fixed effects and random effects models, a discussion that would definitely benefit the presenter mentioned at the outset of this introduction.

—*Tim Futing Liao*
Series Editor

CHAPTER 1. INTRODUCTION

For many years, the most challenging project in statistics has been the effort to devise methods for making valid causal inferences from nonexperimental data. And within that project, the most difficult problem is how to statistically control for variables that cannot be observed. For experimentalists, the solution to that problem is easy: random assignment. Random assignment to treatment groups makes those groups approximately equal on *all* characteristics of the subjects, whether those characteristics are observable or unobservable. But in nonexperimental research, the classic way to control for potentially confounding variables is to measure them and put them in some kind of regression model. No measurement, no control.

In this book, I describe a class of regression methods called fixed effects models that make it possible to control for variables that have not or cannot be measured. The basic idea is very simple: Use each individual as his or her own control. For example, if you want to know whether marriage reduces recidivism among chronic offenders, compare an individual's arrest rate when he is married with his arrest rate when he is not married. Assuming that nothing else changes (a big assumption), the difference in arrest rates between the two periods is an estimate of the marriage effect for that individual. And if we average those differences across all persons in the population, we get an estimate of the "average treatment effect." This estimate controls for all stable characteristics of the offender. It controls both for easily measured variables such as sex, race, ethnicity, and region of birth, and for more difficult variables such as intelligence, parents' child-rearing practices, and genetic makeup. While it does not control for time-varying variables such as employment status and income, these may be handled by the more conventional approach of measuring and putting them in a regression model.

Here's another example. Suppose you want to know whether time spent playing video games affects children's school performance. For a sample of children, you have measures of each variable at several points in time. A basic fixed effects analysis could be accomplished by estimating a separate regression of school performance on video-playing time for each child, then averaging the parameter estimates across children. Because only within-child variation is used to estimate the regression parameters, all stable characteristics of children are controlled.

There are two basic data requirements for using fixed effects methods. First, the dependent variable must be measured for each individual on at least two occasions. Those measurements must be directly comparable, that is, they must have the same meaning and metric. Second, the predictor variables of interest must change in value across those multiple occasions for some substantial portion of the sample. Fixed effects methods are pretty much useless for estimating the effects of variables that don't change over time, such as race and sex. Of course, some statisticians argue that it makes no sense to talk about causal effects of such variables anyway (Sobel, 1995).

Why is a whole book needed for fixed effects methods? First, rather different methods are needed for different kinds of dependent variables, whether quantitative, categorical, count, or event time. Second, for a specific kind of dependent variable, there are often two or more ways to implement the fixed effects approach, and we need to understand their differences and similarities. Third, and most challenging, special methods are needed (but not always available) when measured predictor variables are not "strictly exogenous"—for example, when a dependent variable at one point in time may affect a predictor variable at a later point in time.

The term *fixed effects model* is usually contrasted with *random effects model*. Unfortunately, this terminology is the cause of much misunderstanding and confusion. In the classic view, a fixed effects model treats unobserved differences between individuals as a set of fixed parameters that can either be directly estimated or partialed out of the estimating equations. In a random effects model, unobserved differences are treated as random variables with a specified probability distribution.

If you consult the experimental design literature for explanations of the difference, you will find statements such as the following:

> Common practice is to regard the treatment effects as fixed if those treatment levels used are the only ones about which inferences are sought. . . . If inferences are sought about a broader collection of treatment effects than those used in the experiment, or if the treatment levels are not selected purposefully . . . , it is common practice to regard the treatment effects as random. (LaMotte, 1983, pp. 138–139)

Such characterizations are very unhelpful in a nonexperimental setting, however, because they suggest that a random effects approach is nearly always preferable. Nothing could be further from the truth.

In a more modern framework (Wooldridge, 2002), the unobserved differences are always regarded as random variables. Then, what distinguishes the two approaches is the structure of the associations between the observed variables and the unobserved variables. In a random effects model, the unobserved variables are assumed to be uncorrelated with (or, more strongly, statistically independent of) all the observed variables. In a fixed effects model, the unobserved variables are allowed to have any associations whatever with the observed variables (which turns out to be equivalent to treating the unobserved variables as fixed parameters). Unless you allow for such associations, you haven't really controlled for the effects of the unobserved variables. This is what makes the fixed effects approach so attractive.

Nevertheless, there are also some potentially serious disadvantages of a fixed effects approach. As already noted, a classic fixed effects approach will not produce any estimates of the effects of variables that don't change over time. In this book, we will see some methods that yield estimates of the effects of variables such as sex and race, but those estimates do not really control for unobservables.

Second, in many cases, fixed effects estimates may have substantially larger standard errors than random effects estimates, leading to higher p values and wider confidence intervals. The reason is simple. Random effects estimates use information both within and between individuals. Fixed effects estimates, on the other hand, use only within-individual differences, essentially discarding any information about differences between individuals. If predictor variables vary greatly across individuals but have little variation over time for each individual, then fixed effects estimates will be very imprecise.

Consider the difficulty of estimating the effect of education (years of schooling) on wages by a fixed effects approach. Although years of schooling can certainly change, most people don't start earning significant wages until after their schooling is completed. Some may go on to get additional education after entering the labor force, but the within-person variation will typically be very small relative to the between-person variation. Furthermore, those whose education level does change over the adult years may be quite unlike those whose education level does not change.

Why then would we want to discard the between-individual variation? This is because it is likely to be confounded with unobserved characteristics of the individuals. The idea is to get rid of the "contaminated" variation and use only the variation that produces approximately unbiased estimates of the parameters of interest. So, in statistical terms, we sacrifice efficiency in order to reduce bias. In nonexperimental studies, I think this is often a good trade-off. Keep in mind, however, that fixed effects methods cannot control for unobserved variables that change over time. In studying the effect of marriage on recidivism, for example, it's quite possible that getting married could be associated with increases in income. Unless income is explicitly brought into the regression model, the estimated effect of marriage could actually represent the effect of income.

Interestingly, fixed effects methods are often used in randomized experiments to increase efficiency (i.e., reduce sampling variability) rather than to reduce bias. In the crossover design (Senn, 1993), each subject receives two or more different treatments at different times, with treatments presented in a randomly chosen order. The dependent variable is measured multiple times, at least once for each treatment, and statistical analysis is carried out using only within-subject variation. In this design, bias is not a concern because all subjects get the same treatments, and those treatments are presented in a random order. As a result, there should be virtually no correlation between the treatment and unobserved differences across subjects. Furthermore, by design, all the variation on the predictor variable (the treatment) is within subject, not between subjects, so no information is lost in discarding the between-subject variation. In fact, a fixed effects analysis will typically produce optimally low standard errors because the between-subjects variation is not part of the error term.

Another attractive thing about fixed effects methods is that software for implementing them is already widely available. For the basic linear models in Chapter 2, for example, ordinary least squares regression packages will usually suffice. The advanced linear models of Chapter 6 can be estimated with many programs for doing structural equation modeling. For the logistic regression models in Chapter 3, you can get by with a conventional logistic regression program for the two-period case. The multiperiod case can be handled by a conditional logit program, which is available in most of the more comprehensive statistical packages. Fixed effects models for count data (Chapter 4) can be estimated with conventional Poisson and negative binomial regression packages. And finally, the event history models of Chapter 5 can be estimated with a standard program for doing Cox regression, or (in the case of nonrepeated events) with a conditional logit program.

To optimally benefit from this book, you should also be familiar with the basic principles of statistical inference, including standard errors, confidence intervals, hypothesis tests, p values, bias, efficiency, and so on (Lewis-Beck, 1995). For each of the chapters, you should have some familiarity with the particular regression method on which the fixed effects methods are based. These methods include linear regression for Chapter 2 (Allison, 1999b), logistic regression for Chapter 3 (Allison, 1999a; Pampel, 2000), Poisson and negative binomial regression for Chapter 4 (Dunteman & Ho, 2005), Cox regression for Chapter 5 (Allison, 1984), and linear structural equation modeling for Chapter 6 (Long, 1983).

To do the computations for the examples in Chapters 2 to 5, I have relied on the Stata package (www.stata.com), which has a rather comprehensive set of commands for doing fixed effects regression. Stata programs for all the examples can be found in Appendix 1. Chapter 6 uses the Mplus package (www.statmodel.com), and programs for that chapter can be found in Appendix 2. Portions of this book are derived from the author's earlier SAS® Press publication *Fixed Effects Regression Methods for Longitudinal Data Using SAS®*, Copyright © 2005, SAS Institute Inc. Readers who wish to see how to apply fixed effects methods using SAS should consult that book.

CHAPTER 2. LINEAR FIXED EFFECTS MODELS

Basics

In this chapter, we consider fixed effects methods for data in which the dependent variable is measured on an interval scale and is linearly dependent on a set of predictor variables. We have a set of individuals ($i = 1, \ldots, n$), each of whom is measured at two or more points in time ($t = 1, \ldots, T$). Each time point will be referred to as a "period."

Here's the notation. We let y_{it} be the dependent variable. We have a set of predictor variables that vary over time, represented by the vector $\mathbf{x}_{it}$, and another set of predictor variables $\mathbf{z}_i$ that do not vary over time. (If you're not comfortable with vectors, you can interpret these as single variables). Our basic model for y is

$$y_{it} = \mu_t + \beta \mathbf{x}_{it} + \gamma \mathbf{z}_i + \alpha_i + \varepsilon_{it} \tag{2.1}$$

where μ_t is an intercept that may be different for each period, and β and γ are vectors of coefficients. Although Equation 2.1 seems to be strictly cross-sectional, there is nothing to prevent the $\mathbf{x}_{it}$ vector from including lagged versions of the x variables, except that one must have at least three periods to estimate a model with a lag of one period.

The two "error" terms, α_i and ε_{it}, behave somewhat differently from each other. There is a different ε_{it} for each individual at each point in time, but α_i only varies across individuals, not over time. We regard α_i as representing the combined effect on y of all unobserved variables that are constant over time. On the other hand, ε_{it} represents purely random variation at each point in time.

At this point, I'll make some rather strong assumptions about ε_{it}, namely, that each ε_{it} has a mean of zero, has a constant variance (for all i and t), and is statistically independent of everything else (except for y). The assumption of zero mean is not critical as it is only relevant for estimating the intercept. The constant variance assumption can sometimes be relaxed to allow for different variances for different t. Note that the ε_{it} at any one period is independent of $\mathbf{x}_{it}$ at any *other* period, which means that $\mathbf{x}_{it}$ is *strictly exogenous*. This assumption may be relaxed in some situations, but the issues involved are neither trivial nor purely technical. I'll discuss some of those issues in Chapter 6.

As for α_i, the traditional approach in fixed effects analysis is to assume that this term represents a set of n fixed parameters that can either be directly estimated or removed in some way from the estimating equations. As noted in Chapter 1, we'll take a more modern approach in this chapter by assuming that α_i represents a set of random variables. Although we'll assume statistical independence of α_i and ε_{it}, we allow for *any* correlations between α_i and $\mathbf{x}_{it}$, the vector of time-varying predictors. And if we are not interested in γ, we can also allow for any correlations between α_i and $\mathbf{z}_i$. The inclusion of such correlations distinguishes the fixed effects approach from a random effects approach and allows us to say that the fixed effects method "controls" for time-invariant unobservables. At this point, we don't need to impose any restrictions on the mean and variance of α_i.

The Two-Period Case

Estimation of the model in Equation 2.1 is particularly easy when the variables are observed at only two periods $(T = 2)$. The two equations are then

$$y_{i1} = \mu_1 + \beta\mathbf{x}_{i1} + \gamma\mathbf{z}_i + \alpha_i + \varepsilon_{i1}$$

$$y_{i2} = \mu_2 + \beta\mathbf{x}_{i2} + \gamma\mathbf{z}_i + \alpha_i + \varepsilon_{i2} \qquad (2.2)$$

If we subtract the first equation from the second, we get the "first difference" equation:

$$y_{i2} - y_{i1} = (\mu_2 - \mu_1) + \beta(\mathbf{x}_{i2} - \mathbf{x}_{i1}) + (\varepsilon_{i2} - \varepsilon_{i1}) \qquad (2.3)$$

which can be rewritten as

$$\Delta y_i = \Delta\mu + \beta\Delta\mathbf{x}_i + \Delta\varepsilon_i \qquad (2.4)$$

where Δ indicates a difference score. Note that both α_i and $\gamma\mathbf{z}_i$ have been "differenced out" of the equation. Hence, we no longer have to be concerned about α_i and its possible correlation with $\Delta\mathbf{x}_i$. On the other hand, we also lose the possibility of estimating γ. Since $\mathbf{x}_{i1}$ and $\mathbf{x}_{i2}$ are each independent of ε_{i1} and ε_{i2}, it follows that $\Delta\mathbf{x}_i$ is independent of $\Delta\varepsilon_i$. This implies that one can get unbiased estimates of β by doing ordinary least squares (OLS) regression on the difference scores.

We now apply this method to some real data. Our sample comes from the National Longitudinal Survey of Youth (NLSY; Center for Human Resource Research, 2002).[1] Drawing a subset of a much larger sample, we have 581 children who were interviewed in 1990, 1992, and 1994. Initially, we will work with just three variables that were measured in each of the three interviews:

ANTI   antisocial behavior (scale ranges from 0 to 6)

SELF   self-esteem (scale ranges from 6 to 24)

POV   coded 1 if family is in poverty, otherwise 0

At this point, we're going to ignore the observations in the middle year, 1992, and use only the data in 1990 and 1994. Our objective is to estimate a linear equation with ANTI as the dependent variable[2] and SELF and POV as independent variables:

$$\text{ANTI}_t = \mu_t + \beta_1 \text{SELF}_t + \beta_2 \text{POV}_t + \alpha + \varepsilon_t, \quad t = 1, 2 \qquad (2.5)$$

By expressing the model in this way, we are assuming a particular direction of causation, specifically, that SELF and POV affect ANTI and not the reverse. We also assume that the effects are contemporaneous (no lagged effects of SELF and POV). Both these assumptions will be relaxed in Chapter 6. Last, we assume that β_1 and β_2 are the same at both periods, an assumption that we will relax very shortly. On the other hand, we let the intercept μ_t be different at each period, allowing for changes in the average level of antisocial behavior that are not a consequence of any changes in SELF or POV.

Let us begin by estimating Equation 2.5 separately for each period, using ordinary least squares regression. The results are shown in the first two columns of Table 2.1. Not surprisingly, in both years, poverty is associated with higher levels of antisocial behavior, whereas self-esteem is associated with lower levels. The coefficients are quite similar across the two years.

In neither of these two regressions are there any controls for time-invariant predictors, such as sex and race. Rather than putting such variables in the equation, however, we can control for *all* time-invariant predictors by doing the regression with difference scores. For each child and each variable, we subtract the 1990 value from the 1994 value and then regress the ANTI difference on the SELF difference and the POV difference. Since POV is a dummy variable, it might seem inappropriate to subtract one value from the other. But, in fact, dummy variables can be treated just like any other variables in this regard.

The results are in the last column of Table 2.1. Although the equation was estimated in the form of difference scores, the coefficients can be interpreted as if we had estimated Equation 2.5 directly. That is, they represent the effects of each variable in a given year on the value of the dependent variable in the

Table 2.1   OLS Regression of Antisocial Behavior on Self-Esteem and
Poverty

	1990		1994		Difference Score	
	Coefficient	Standard Error	Coefficient	Standard Error	Coefficient	Standard Error
Intercept	2.375**	0.384	2.888**	0.447	0.209**	0.063
Self-esteem	−0.050**	0.019	−0.064**	0.021	−0.056**	0.015
Poverty	0.595**	0.126	0.547**	0.148	−0.036	0.128
R^2	0.05		0.04		0.02	

**$p < .01$.

same year. For self-esteem, the estimated coefficient in the difference score
model is in between the coefficients for the two separate years and still highly
significant. For poverty, on the other hand, the coefficient is dramatically
smaller and no longer statistically significant.

It's fairly common for fixed estimates to vary markedly from those
produced by other methods. In this case, one possible explanation is that
the estimates for the poverty effect in the regressions for the separate
years were spurious, reflecting the correlation between poverty and some
time-invariant variables that affected antisocial behavior.

One should not be too hasty about that conclusion, however. Whenever
conventional regression produces a significant coefficient but fixed effects
regression does not, there are two possible explanations: (a) The fixed
effects coefficient is substantially smaller in magnitude and/or (b) the fixed
effects standard error is substantially larger. As already mentioned, standard
errors for fixed effects coefficients are often larger than those for other
methods, especially when the predictor variable has little variation over time.
In fact, most of the variation in poverty is between girls, with only about 24%
of the girls moving into or out of poverty between 1990 and 1994.

Nevertheless, the standard error of the poverty coefficient in the difference
score model is about the same as the standard error in 1990 and smaller than
the standard error in 1994. The conclusion, then, is that insufficient variation
is not a problem here. There seems to be a real and substantial decline in the
magnitude of the poverty effect when time-invariant variables are controlled.
The general lesson is this: Whenever p values for fixed effects methods are

noticeably different from those for other methods, always check both the coefficients and their standard errors.

Note, finally, that the intercept of 0.209 is highly significant. This coefficient represents the estimated change in antisocial behavior from Time 1 to Time 2 for a person who did *not* change in self-esteem or poverty.

Extending the Difference Score Method for the Two-Period Case

The basic fixed effects model of Equation 2.1 can be extended to allow the effects of **x** and **z** to vary over time. In the two-period case, we can write the equations with distinct coefficients at each period:

$$y_{i1} = \mu_1 + \beta_1 \mathbf{x}_{i1} + \gamma_1 \mathbf{z}_i + \alpha_i + \varepsilon_{i1}$$

$$y_{i2} = \mu_2 + \beta_2 \mathbf{x}_{i2} + \gamma_2 \mathbf{z}_i + \alpha_i + \varepsilon_{i2} \tag{2.6}$$

Taking first differences and rearranging terms produces

$$y_{i2} - y_{i1} = (\mu_2 - \mu_1) + \beta_2(\mathbf{x}_{i2} - \mathbf{x}_{i1}) + (\beta_2 - \beta_1)\mathbf{x}_{i1} \\ + (\gamma_2 - \gamma_1)\mathbf{z}_i + (\varepsilon_{i2} - \varepsilon_{i1}) \tag{2.7}$$

which could also be written as

$$\Delta y_i = \Delta \mu + \beta_2 \Delta \mathbf{x}_i + \Delta\beta \mathbf{x}_1 + \Delta\gamma \mathbf{z}_i + \Delta\varepsilon_i$$

There are three things to note about this equation. First, as before, α_i has dropped out, so we don't have to be concerned about its potential confounding effects. Second, **z** has *not* dropped out, and its coefficient vector is the difference in the coefficient vectors for the two time points. From this we learn that time-invariant variables whose coefficients vary over time *must* be explicitly included in the regression equation. Fixed effects only controls for time-invariant variables with time-invariant effects. Third, the equation now includes $\mathbf{x}_1$ as a predictor, and its coefficient vector is the difference in the coefficient vectors for the two time periods. Thus, for **z** and $\mathbf{x}_1$, tests for whether their coefficients are 0 are equivalent to testing whether $\beta_1 = \beta_2$ or $\gamma_1 = \gamma_2$.

Let's try this for the NLSY data. The data set includes several time-invariant variables that we'll examine as possible predictors:

BLACK	1 if child is black, otherwise 0
HISPANIC	1 if child is Hispanic, otherwise 0

CHILDAGE	child's age in 1990
MARRIED	1 if mother was currently married in 1990, otherwise 0
GENDER	1 if female, 0 if male
MOMAGE	mother's age at birth of child
MOMWORK	1 if mother was employed in 1990, otherwise 0

The first two variables, BLACK and HISPANIC, represent two categories of a three-category variable—the reference category being white, non-Hispanic. These seven variables will now be included in the difference score regression for antisocial behavior, along with the difference scores for self-esteem and poverty. The 1990 measures of self-esteem and poverty are also included. Results shown in Table 2.2 are consistent with the findings in Table 2.1. The coefficient for self-esteem (the difference score) is about −.06 and is highly significant, while the coefficient for poverty (the difference score) is

Table 2.2   OLS Estimates of Extended Difference Score Model

	Coefficient	Standard Error	p Value
Intercept	−0.550	1.360	.6859
Self-esteem difference	−0.060	0.020	.0024
Poverty difference	0.031	0.156	.8446
Self-esteem 1990	−0.018	0.025	.4826
Poverty 1990	0.121	0.178	.4991
Black	−0.100	0.155	.5185
Hispanic	0.084	0.164	.6109
ChildAge	0.220	0.107	.0409
Married	−0.206	0.154	.1808
Gender	0.101	0.126	.4262
MomAge	−0.040	0.030	.1842
MomWork	−0.153	0.140	.2735

far from significant. Neither the 1990 self-esteem coefficient nor the 1990 poverty coefficient approaches statistical significance, indicating that the effects of self-esteem and poverty were stable over time. Of the seven time-invariant variables, only one—child's age in 1990—is statistically significant (just barely). That doesn't mean that the other six variables are not affecting antisocial behavior. Rather, it means that their effects in 1990 and 1994 are essentially the same.

A First-Difference Method for Three or More Periods per Individual

When each individual is observed at three or more points in time ($T > 2$), it's not so obvious how to extend the methods we just considered. For the NLSY data, we actually have three years of data—1990, 1992, and 1994. One possible approach is to construct and estimate two first-difference equations. Starting with Equation 2.2, we have

$$y_{i2} - y_{i1} = (\mu_2 - \mu_1) + \beta(\mathbf{x}_{i2} - \mathbf{x}_{i1}) + (\varepsilon_{i2} - \varepsilon_{i1})$$

$$y_{i3} - y_{i2} = (\mu_3 - \mu_2) + \beta(\mathbf{x}_{i3} - \mathbf{x}_{i2}) + (\varepsilon_{i3} - \varepsilon_{i2}) \qquad (2.8)$$

These equations can be estimated separately by OLS, and each will give unbiased estimates of β. The first two columns of Table 2.3 show the results for the NLSY data. The negative coefficients for self-esteem are highly significant in both difference equations and are roughly of the same magnitude. Poverty is far from significant in both equations. The intercepts represent the change from one period to the next, after adjusting for the two predictor variables. Although there was an increase in antisocial behavior for each 2-year interval, only the change from 1992 to 1994 was statistically significant.

Under the assumption that β doesn't vary over time, the two equations should be estimated simultaneously for optimal efficiency. This can be accomplished by creating a single data set with two records for each person, one with the difference scores for the first equation and the other with the difference scores for the second equation. There should also be a dummy variable distinguishing the first record from the second record. And also, there should be a variable with a common ID number for the two records from each person.

The third column of Table 2.3 shows the results of applying OLS to this combined data set of 1,162 records. Not surprisingly, the coefficients for self-esteem and poverty are in between their respective coefficients in the first two columns. The standard errors are appreciably lower, however,

Table 2.3 First-Difference Regressions of Antisocial Behavior on Self-Esteem and Poverty

| | 1992–1994 OLS | | 1990–1992 OLS | | Combined OLS | | Combined GLS | |
	Coefficient	Standard Error	Coefficient	Standard Error	Coefficient	Standard Error	Coefficient	Standard Error
Intercept	0.71**	0.059	0.040	0.053	0.045	0.056	0.05	0.056
Self-esteem	–0.072**	0.016	–0.039**	0.014	–0.055**	0.010	–0.055**	0.010
Poverty	0.216	0.136	0.197	0.133	0.213	0.095	0.139	0.094
Equation dummy					0.122	0.080	0.122	0.094

$**p < .01$.

because more information is being used. The intercept can be interpreted as an estimate of $\mu_2 - \mu_1$ while the coefficient for the equation dummy is an estimate of $(\mu_3 - \mu_2) - (\mu_2 - \mu_1)$. Both are positive, indicating an increase in antisocial behavior from Time 1 to Time 2, and a more rapid increase from Time 2 to Time 3. But neither is statistically significant.

Although the combined OLS estimates should be unbiased, they ignore the fact that $\varepsilon_2 - \varepsilon_1$ is likely to be negatively correlated with $\varepsilon_3 - \varepsilon_2$ because they share a common component, ε_2, with opposite signs. This implies that the coefficient estimates may not be fully efficient and the standard error estimates may be biased. We can solve this problem by estimating the correlation between the error terms and then using generalized least squares (GLS) to take account of that correlation.

Most comprehensive statistical packages have routines to do GLS. Such programs typically require the specification of an ID variable so that observations from the same individual can be identified. I used the **xtreg** command in Stata with the **pa** option, which estimates the linear model using GLS.[3] The GLS results, shown in the last column of Table 2.3, are very similar to the OLS coefficient estimates and standard errors in the preceding column.

The first-difference method can be easily extended to more than three periods per individual. For T periods per individual, $T - 1$ records are created, each with difference scores between adjacent time points for all variables. Additionally, there should be a variable containing a common ID number for all observations from the same individual, and a variable or a set of dummy variables to distinguish the different records. The regression is then estimated on the entire set of records, using GLS to adjust for correlations among the error terms. Unless T is large, for example, greater than 10, it's probably best to allow the error correlation matrix to be unstructured. That is, the matrix would allow for a different correlation between each pair of error terms. With larger T, it may be preferable to impose a simplified structure to reduce the number of distinct correlations that need to be estimated. For more details, see Greene (2000).

Dummy Variable Method for Two or More Periods per Individual

Although the multiple-difference-score method is a reasonable way to estimate a fixed effects model for the multiperiod case, the name "fixed effects" is usually reserved for a different method, one that can be implemented either by dummy variables or by constructing mean deviations. The results produced by the fixed effects method are not identical to those produced by the difference-score method, although they will usually be very similar. In the two-period case, the two methods give identical results.

The dummy variable method requires a data set with a rather different structure: one record for each period for each individual. For the NLSY data, for example, the required data set has three records for each of the 581 children, for a total of 1,743 records. The time-varying variables have the same variable names on each record but different values. For any time-invariant variables, their values are simply replicated across the multiple records for each individual. There should also be an ID variable with a common value for all the records for each individual. Last, there should be a variable distinguishing the different periods for each individual. For the NLSY data, for example, the variable TIME has values of 1, 2, and 3, corresponding to 1990, 1992, and 1994. Table 2.4 shows the first 15 records of this data set, corresponding to the first five persons.

Table 2.4 Data Set With Three Observations per Person (First Five Persons)

ID	TIME	ANTI	SELF	POV	GENDER
1	1	1	21	1	1
1	2	1	24	1	1
1	3	1	23	1	1
2	1	0	20	0	1
2	2	0	24	0	1
2	3	0	24	0	1
3	1	5	21	0	0
3	2	5	24	0	0
3	3	5	24	0	0
4	1	2	23	0	0
4	2	3	21	0	0
4	3	1	21	0	0
5	1	1	22	0	1
5	2	0	23	0	1
5	3	0	24	0	1

To implement the method, it's necessary to construct a set of dummy variables to distinguish the individuals in the data set. In our example, that means 580 dummy variables to represent the 581 children. Many statistical packages can do this automatically by specifying the ID variable as a categorical variable. If the TIME variable is also specified as categorical, two dummy variables will be created to distinguish the three different years. One can then use OLS to estimate the coefficients. The coefficients for the dummy variables created from the ID variable are actually the estimates of the α_i in Equation 2.1, under the constraint that one of them is equal to 0.

I did this in Stata using the **reg** command,[4] with results shown in the left-hand panel of Table 2.5. Only the coefficients for the first nine dummy variables are shown.

Table 2.5 Regression of Antisocial Behavior on Self-Esteem and Poverty, Dummy Variable Method

	Fixed Effects			Conventional OLS		
	Coefficient	Standard Error	p	Coefficient	Standard Error	p
SELF	−0.055	0.010	.00	−0.067	0.011	.00
POV	0.112	0.093	.23	0.518	0.079	.00
TIME_2	0.044	0.059	.45	0.051	0.090	.58
TIME_3	0.211	0.059	.00	0.223	0.091	.01
ID_2	−0.887	0.819	.28			
ID_3	4.131	0.811	.00			
ID_4	1.057	0.819	.20			
ID_5	−0.536	0.819	.51			
ID_6	0.040	0.820	.96			
ID_7	2.170	0.821	.01			
ID_8	0.910	0.820	.27			
ID_9	−0.276	0.819	.74			

Comparing the results in Table 2.5 with those in the last column of Table 2.3 (obtained with the first-difference method), we see that the coefficient and standard error for self-esteem are virtually identical. The coefficient for poverty is a little smaller using the dummy variable method, but it is far from significant with either method. The coefficients for TIME_2 and TIME_3 represent contrasts with the omitted category (TIME_1). We see that antisocial behavior increases, on average, over time, and that TIME_3 is significantly higher than TIME_1.

For comparison, the right-hand panel of Table 2.5 gives the OLS estimates of the coefficients without the inclusion of the 580 dummy variables. As we saw in the two-period case, the big difference in the results for the two methods is that the coefficient for POV is much larger for conventional OLS, and highly significant. Thus, the apparent effect of poverty on self-esteem disappears when we adjust for all between-person differences and focus only on within-person changes. It's also of some interest to compare the standard errors. The standard error for the POV coefficient is larger for the fixed effects estimate, a fairly typical result that stems from not using the between-person variation. On the other hand, for the coefficients for SELF and the two TIME dummies, the fixed effects standard errors are actually smaller than those for conventional OLS. Why the difference? It's all a matter of the relative magnitudes of within- and between-person variation. For POV, 70% of the variation is between persons, while for SELF the figure is only 53%.[5] For the TIME dummies, all the variation is within person and none is between. The best situation for a fixed effects analysis is when all the variation on a time-varying predictor is within persons, but there's still a lot of between-person variation on the response variable.

The problem with the dummy variable method is that the computational requirement of estimating coefficients for all the dummy variables can be quite burdensome, especially in large samples where it may be beyond the capacity of the software or the machine memory. Fortunately, there is an alternative algorithm—the mean deviation method—that produces exactly the same results. The one drawback is that it doesn't give estimates for the coefficients of the dummy variables representing different persons, but those are rarely of interest anyway.

The mean deviation algorithm works like this. For each person and for each time-varying variable (both response and predictor variables), we compute the means over time for that person:

$$\bar{y}_i = \frac{1}{n_i} \sum_T y_{it}$$

$$\bar{\mathbf{x}}_i = \frac{1}{n_i} \sum_T \mathbf{x}_{it}$$

where n_i is the number of measurements for person i. Then we subtract the person-specific means from the observed values of each variable:

$$y^*_{it} = y_{it} - \bar{y}_i$$

$$\mathbf{x}^*_{it} = \mathbf{x}_{it} - \bar{\mathbf{x}}_i$$

Finally, we regress y^* on $\mathbf{x}^*$, plus variables to represent the effect of time. This is sometimes called a "conditional" method because it conditions out the coefficients for the fixed effects dummy variables.

If you construct the deviation scores yourself and then use an ordinary regression program to estimate the coefficients, you will get the correct OLS estimates for all the coefficients. However, the standard errors and p values will not be correct. That's because the calculation of the degrees of freedom is based on the number of variables in the specified model, when it should actually include the number of dummy variables implicitly used to represent different persons in the sample (580 for the NLSY data). Formulas are available to correct the standard errors and p values (Judge, Hill, Griffiths, & Lee, 1985), but it's much easier to let the software do it for you. For example, the **xtreg** command[6] in Stata does the correct calculations for a fixed effects model; SAS does it with the ABSORB statement in PROC GLM.

Using the **xtreg** command, I specified a fixed effects model (FE option) with ID as the variable that identifies records from the same person. Results were identical to the first five lines of Table 2.5. **xtreg** also reports several additional statistics that are specific to fixed effects models:

1. An F test of the null hypothesis that all the coefficients for the fixed effects dummy variables are zero. In this case, the p value is less than .0001, so the null hypothesis can be confidently rejected. This is equivalent to saying that there is evidence for person-level unobserved heterogeneity. That is, there are stable differences in antisocial behavior between persons that are not fully accounted for by the measured predictor variables.

2. An estimate of the proportion of variance in the dependent variable that is attributable to the fixed effects (the α_is), labeled "rho (fraction of variance due to u_i)." In this case, the estimate is 0.64.

3. An estimate of the correlation between the fixed effects α_i and $\hat{\beta}\mathbf{x}_{it}$, the estimated linear combination of the time-varying predictors. In a random effects model, this correlation is assumed to be 0. For these data, the correlation is .068.

4. Three R^2s: within, between, and overall. The within R^2 is just the usual R^2 calculated for the regression using the mean deviation variables. Here it's .033. The between R^2 is the squared correlation between the person-specific mean of y and the *predicted* person-specific mean of y. In this case, it's .041. Finally, the overall R^2 (.036 in this case) is the squared correlation between y itself and the predicted value of y. All three of these R^2s are calculated using predicted values based on the estimated regression coefficients but *not* using the coefficients for the fixed effects dummy variables. If you include those (using the dummy variable method), the R^2 goes up to .73 for these data.

As noted before, one characteristic of this method is the inability to estimate coefficients for time-invariant predictors. This is evident from the fact that subtracting the person-specific mean of a time-invariant predictor from the individual values (which are the same at all periods) yields a value of 0 for all persons. Keep in mind, however, that we are still controlling all time-invariant predictors even though they drop out of the equation. In the next section, moreover, we'll see how to test whether the effects of those variables are themselves time invariant.

Interactions With Time
in the Fixed Effects Method

In the two-period case, we saw how to extend the method to allow the coefficients for predictor variables to change over time. For a time-varying predictor, we simply added the variable measured at time 1 to the model. For a time-invariant predictor, we added the variable itself to the model. For the dummy variable method (or the equivalent mean deviation method), these extensions are accomplished by including interactions with time.

For the three-period NLSY data, Table 2.6 shows the results of including interactions between TIME (treated as categorical) and both time-varying and time-invariant predictors. Since TIME has three categories, there are two interactions with each predictor. Note that the time-invariant predictors do not have main effects included in the model. If we had tried to include them, the software would have dropped them from the model because they have no variation within persons.

For each of the interactions, the t statistic tests whether a coefficient at Time 2 or Time 3 is different from the coefficient at Time 1. Of the 18 interactions, only one (TIME_3*CHILDAGE) is statistically significant ($p = .024$). For that interaction, the coefficient of 0.227 indicates that the

Table 2.6 Interactions With Time

	Coefficient	Standard Error	t	p
TIME_2	0.291	1.245	0.23	.82
TIME_3	−0.444	1.258	−0.35	.72
SELF	−0.034	0.016	−2.08	.04
POV	0.097	0.130	0.75	.46
TIME_2 * SELF	−0.026	0.020	−1.28	.20
TIME_3 * SELF	−0.023	0.021	−1.09	.28
TIME_2 * POV	−0.112	0.152	−0.74	.46
TIME_3 * POV	0.099	0.155	0.64	.52
TIME_2 * BLACK	0.250	0.144	1.74	.08
TIME_3 * BLACK	−0.110	0.144	−0.77	.44
TIME_2 * HISPANIC	0.190	0.154	1.23	.22
TIME_3 * HISPANIC	0.075	0.153	0.49	.62
TIME_2 * CHILDAGE	0.076	0.100	0.76	.45
TIME_3 * CHILDAGE	0.227	0.100	2.26	.02
TIME_2 * MARRIED	−0.095	0.143	−0.67	.51
TIME_3 * MARRIED	−0.176	0.143	−1.23	.22
TIME_2 * GENDER	0.041	0.118	0.35	.73
TIME_3 * GENDER	0.107	0.118	0.91	.37
TIME_2 * MOMAGE	−0.027	0.028	−0.96	.34
TIME_3 * MOMAGE	−0.042	0.028	−1.52	.13
TIME_2 * MOMWORK	0.0137	0.131	1.05	.29
TIME_3 * $MOMWORK	−0.144	0.130	−1.11	.27

coefficient for CHILDAGE is 0.227 higher at Time 3 than at Time 1. Of course, with 18 tests it's a pretty good bet that at least one of them is statistically significant at the .05 level even if there's nothing really going on. A simultaneous test that all 18 interaction coefficients are equal to 0 yields a p value of .15.

Comparison With Random Effects Models

A popular alternative to the linear fixed effects model is the random effects or mixed model. This model is based on the same equation that we used for the fixed effects model:

$$y_{it} = \mu_t + \beta \mathbf{x}_{it} + \gamma \mathbf{z}_i + \alpha_i + \varepsilon_{it} \qquad (2.9)$$

The crucial difference is that now, instead of treating α_i as a set of fixed numbers (which is equivalent to treating α_i as random but with all possible correlations with $\mathbf{x}_{it}$), we assume that α_i is a set of random variables with a specified probability distribution. For example, it is typical to assume that each α_i is normally distributed with a mean of 0, constant variance, and is independent of all the other variables on the right-hand side of the equation.

There's a lot of software available to estimate the random effects model. SAS can do it with the MIXED procedure. The **xtreg** command in Stata does GLS estimation of the random effects model by default. Table 2.7 presents the estimates produced by **xtreg**, both with and without time-invariant predictors.

The fact that the random effects method can include time-invariant predictors is the most apparent difference between the fixed and the random effects models. In this case, however, we see that the inclusion of those variables doesn't make much difference in the coefficients for the time-varying predictors, self-esteem, and poverty.

Like the conventional OLS estimates in Table 2.5 and unlike the fixed effects estimates, both variables have coefficients that are highly significant. The similarity of random effects estimates and OLS estimates is not surprising. If the random effects assumption that α is uncorrelated with all other variables is correct, both methods produce consistent (and therefore approximately unbiased estimates) of the coefficients in Equation 2.9. But when that assumption is incorrect, both methods will yield biased estimates.

Why is it that POV is highly significant in the random effects model but far from significant in the fixed effects model? As explained earlier, whenever a coefficient is significant in a random effects model but not in a

Table 2.7   GLS Estimates of Random Effects Models

	Coefficient	Standard Error	p	Coefficient	Standard Error	p
SELF	–0.062	0.009	.00	–0.060	0.009	.00
POV	0.247	0.080	.00	0.296	0.077	.00
TIME_2	0.047	0.059	.42	0.047	0.059	.42
TIME_3	0.216	0.059	.00	0.216	0.059	.00
BLACK	0.227	0.126	.07			
HISPANIC	–0.218	0.138	.11			
CHILDAGE	0.088	0.091	.33			
MARRIED	–0.049	0.126	.70			
GENDER	–0.483	0.106	.00			
MOMAGE	–0.022	0.025	.39			
MOMWORK	0.261	0.115	.02			

fixed effects model, the first thing to do is compare the standard error estimates. Because fixed effects standard errors are often substantially higher than random effects standard errors, that alone could explain the higher p values. In this case, however, although the fixed effects standard error for POV is a little higher (0.09 vs. 0.08), that's not enough to account for the difference. Even if we substitute 0.08 for 0.09, the POV coefficient would still not be significant in the fixed effects model. Clearly, the main difference is in the magnitude of the two coefficients, 0.11 for fixed effects, and 0.25 (or 0.30) for random effects (depending on whether the time-invariant predictors are controlled or not). The most plausible explanation for this difference is that there are unobserved variables that "explain away" the observed association between poverty and antisocial behavior. When these variables are controlled, via fixed effects, the relationship disappears.

The key point here is that, contrary to popular belief, estimating a random effects model does not really "control" for unobserved heterogeneity. That's because the conventional random effects model assumes no correlation between the unobserved variables and the observed variables. The fixed effects model, on the other hand, allows for any correlations between time-invariant predictors and the time-varying predictors. It does so, however, at the cost of some efficiency in the event that those correlations are really zero.

It has been shown that the random effects model is actually a special case of the fixed effects model (Mundlak, 1978). That is, if you start with the conventional random effects model of Equation 2.9 and then allow for all possible correlations between the α_i term and the $\mathbf{x}_{it}$ variables, you end up with something equivalent to the fixed effects model. In general, whenever there is a choice between two nested models, one being a restricted version of the other, there is a trade-off between bias and efficiency. The simpler model (the random effects model) will lead to more efficient estimates, but those estimates may be biased if the restrictions of the model are wrong. The more complex model (the fixed effects model) is less prone to bias but at the expense of greater sampling variability.

Given these trade-offs, it would be useful to have a statistical test that compared the random effects and fixed effects models. Such a test would help determine whether the biases inherent in the random effects method are small enough to ignore, or whether we need to move to the less restrictive fixed effects model. The best-known test is the Hausman (1978) test of the null hypothesis that the random effects coefficients are identical to the fixed effects coefficients.[7] This test is available in several software packages. For the example at hand, the fairest test would be to compare the fixed effects coefficients in Table 2.5 with the random effects coefficients in the left-hand panel of Table 2.7, which control for several time-invariant variables. When I used this test with the **xtreg** command in Stata, I got a p value of .04, suggesting some evidence against the random effects model and in favor of the fixed effects model. In the next section, I consider an alternative test that may have somewhat better properties than the Hausman test currently implemented in Stata.

A Hybrid Method

We shall now consider an approach that combines some of the virtues of the fixed effects and random effects models. As we saw earlier, one way to estimate the fixed effects model is to express all variables as deviations from their person-specific means, then run an OLS regression on those mean deviation variables. In the hybrid method, the time-varying $\mathbf{x}$ variables are again transformed into deviations from their person-specific means, but the response variable y is not. Furthermore, unlike our previous fixed effects methods, we now include the time-invariant $\mathbf{z}$ variables in the regression model. In addition, we also include variables that are the person-specific means for each of the time-varying variables (which are also time invariant). Finally, instead of doing OLS regression, we estimate a random effects model to ensure that the standard errors reflect the dependence among the multiple observations for each person.[8] Table 2.8 shows the results for the NLSY data. DSELF and DPOV are the deviation variables. MSELF and MPOV are the person-specific means. The first thing to notice in this table is that coefficients and standard errors for DSELF, DPOV, and the two time dummies are *identical* to those we saw for the fixed effects method in Table 2.5. Thus, we have yet another method for producing fixed effects estimates.[9] Actually, for DSELF and DPOV, it doesn't matter what time-invariant variables are in the equation. Even if we delete MSELF, MPOV, and the other time-invariant variables, the coefficients and standard errors for the time-varying variables will remain the same. Of course, what we get from this hybrid approach are estimates for the effects of the time-invariant variables, something that we couldn't obtain from the conventional fixed effects method.

Table 2.8   Hybrid Estimates

	Coefficient	Standard Error	p
DSELF	−0.055	0.010	.00
DPOV	0.112	0.093	.22
MSELF	−0.090	0.022	.00
MPOV	0.616	0.157	.00
BLACK	0.111	0.132	.40
HISPANIC	−0.280	0.139	.04
CHILDAGE	0.086	0.091	.35
MARRIED	−0.128	0.128	.32
GENDER	−0.508	0.107	.00
MOMAGE	−0.011	0.025	.65
MOMWORK	0.164	0.119	.17
TIME_2	0.044	0.059	.45
TIME_3	0.211	0.059	.00

In the multilevel model literature (Bryk & Raudenbusch, 1992; Goldstein, 1987; Kreft & De Leeuw, 1995), the practice of subtracting person-specific means from each time-varying variable is called *group mean centering*. Although it is well-known that group mean centering can produce substantially different results, this literature has not made the connection to fixed effects models, nor has it been recognized that group mean centering controls for all time-invariant predictors.

The estimates for the mean variables, MSELF and MPOV, are not particularly enlightening in themselves. But it's important to include these variables for two reasons. First, they help us get better estimates of the effects of the other time-invariant variables. Excluding MSELF and MPOV would mean that we are not fully controlling for these variables. Second, comparing their coefficients with those of the mean deviation variables, DSELF and DPOV, give us some insight into what's going on here. If the assumptions of the random effects model are correct (i.e., the α_i term is uncorrelated with the **x** variables), then the deviation coefficient should be the same as the mean coefficient for each variable (apart from sampling variability). This isn't far from true for DSELF and MSELF. But the coefficient for MPOV is much larger than the coefficient for DPOV. When we estimated the conventional random effects model, what we got for SELF and POV were weighted averages of these "within" and "between" coefficients. This further suggests that we can test the random effects model against the fixed effects model by testing equality for these two pairs of coefficients (an alternative to the Hausman test discussed earlier). This is easy to do in Stata (using a Wald test) and yields a p value of .007, fairly clear-cut evidence against the random effects model.

Another attraction of the hybrid approach is that it can be extended in interesting ways that are not easily handled with the conventional methods for fixed effects estimation. The random effects models that we have been considering so far are all random *intercept* models. It's also possible, to estimate random *slope* models. For example, instead of assuming that the coefficient for DSELF is the same for everyone, we could assume that it is a random variable, and then estimate its mean and standard deviation. Such models are easily handled with the MIXED procedure in SAS, or with the **xtmixed** command in Stata. The latter produced an estimated mean of −0.055 for the coefficient of DSELF. The estimated standard deviation for the coefficient was 0.070. That's more than twice its standard error of 0.024, which strongly suggests that the effect of DSELF varies across persons.

Using the hybrid method, it's also possible to estimate models with more complex error structures (e.g., three-level structures or autoregressive structures) than the rather simple structure implied by the conventional fixed effects model. For further information on such models, see Singer and Willett (2003).

Summary

There are several equivalent ways to estimate linear fixed effects models for quantitative response variables:

1. When each individual is observed at exactly two periods, compute the difference scores for all time-varying predictors. Then, do OLS regression of the difference score for the response variable on the difference scores for the predictors.

2. For any number of periods, structure the data so that there is one record for each period for each individual. Then do OLS regression with the inclusion of dummy variables for all individuals (excluding one).

3. For the data structure in Method 2, transform all the time-varying variables into deviations from individual-specific means. Then do OLS regression on the deviation scores (with corrections for standard errors test statistics and p values). This is conveniently done in Stata with the **xtreg** command.

4. For the data structure in Method 2, transform only the *predictor* variables into deviations from individual-specific means. Then estimate a random effects model, with predictor variables that include both the means and deviations from the means.

Of these methods, the fourth is the most flexible. It offers the following capabilities not shared with one or more of the other methods:

- The inclusion of predictor variables that do not vary within individuals
- A test of the fixed effects versus random effects assumption
- Random coefficients for those predictors that vary within individuals
- Less restrictive error structures

Regardless of which computational method is used, the fixed effects method effectively controls for all time-invariant predictors, both measured and unmeasured. This is its principal attraction as compared with random effects methods. A key assumption of the fixed effects method, however, is that the time-invariant predictors must have the same effects at all occasions. Variables whose effects are not constant across occasions must be explicitly included in the model. And, of course, the fixed effects approach does nothing to control for unmeasured predictors that vary over time.

Notes

1. I am indebted to Peter Tice for preparing this data set and making it available to me.

2. Because ANTI only takes on integer values and is positively skewed, an argument could be made for estimating ordered logit models rather than linear models. Indeed, we shall estimate such models for these data in Chapter 3. However, the qualitative results of the logit models are nearly the same as those for the linear models in this chapter.

3. A random effects model (the default in **xtreg**) is not appropriate in this case because the model allows only positive correlations between the error term. With multiple first-difference equations, the correlations are often negative.

4. The **reg** command (and all other Stata commands discussed in this chapter) was used with the **xi** prefix in order to treat TIME and ID as categorical variables.

5. These numbers can be obtained by running an analysis of variance with each variable as the dependent variable and the ID variable as a categorical predictor.

6. Another Stata command that implements the mean-deviation method is **areg** with the **absorb(id)** option.

7. The Hausman test is computed as follows. Let **b** be the vector of fixed effects coefficients (excluding the constant) and let β be the corresponding vector of random effects coefficients. Let $\Sigma = \mathrm{var}(\mathbf{b}) - \mathrm{var}(\beta)$, where $\mathrm{var}(\mathbf{b})$ is the estimated covariance matrix for **b** and similarly for β. The statistic is then $m = (b-\beta)' \Sigma^{-1} (b-\beta)$, which has a chi-square distribution under the null hypothesis.

8. The hybrid method described here is similar, but not identical, to methods proposed by Mundlak (1978) and Hausman (1978).

9. The estimates are identical only if the data set is balanced—that is, there is an equal number of observed periods for every individual. Otherwise the hybrid estimates will be slightly different from the conventional fixed effects estimates.

CHAPTER 3. FIXED EFFECTS LOGISTIC MODELS

In this chapter, we shall see how the fixed effects methods of the last chapter can be generalized to allow for categorical response variables. To explore these methods, we'll use a data set that comes from the National Longitudinal Survey of Youth (NLSY). This data set has 1,151 teenage girls who were interviewed annually for 5 years beginning 1979. The response variable POV1-POV5 is a dichotomy: whether or not the girl's household was in poverty according to U.S. federal standards in each of the 5 years, coded 1 for poverty and 0 for nonpoverty. We have the following predictor variables:

AGE	Age in years at the first interview
BLACK	1 if respondent is black, otherwise 0
MOTHER	1 if respondent currently has a least one child, otherwise 0
SPOUSE	1 if respondent is currently living with a spouse, otherwise 0
SCHOOL	1 if respondent is currently enrolled in school, otherwise 0
HOURS	Hours worked during the week of the survey

The first two variables are time invariant, whereas the last four may differ at each interview.

Instead of linear models, we now work with logistic regression models. Analogous to Equation 2.1, our basic model is

$$\log\left(\frac{p_{it}}{1 - p_{it}}\right) = \mu_t + \beta \mathbf{x}_{it} + \gamma \mathbf{z}_i + \alpha_i, \quad t = 1, 2, \dots, T \qquad (3.1)$$

where p_{it} is the probability that the response variable is equal to 1. As before, $\mathbf{x}_{it}$ is a vector of time-varying predictors, $\mathbf{z}_i$ is a vector of time-invariant predictors, and α_i represents the combined effects of all unobserved variables that are constant over time. In this chapter, we shall treat α_i as a set of fixed constants, one for each individual. However, this is equivalent to assuming that α_i is random with unrestricted associations between α_i and $\mathbf{x}_{it}$.

The Two-Period Case

In Chapter 2, we saw that fixed effects linear models for the two-period case could be estimated by calculating difference scores for all variables and then running an ordinary least squares regression. An analogous procedure is available for logistic regression, but there are some important differences.

Table 3.1 Cross-Classification for Poverty in Years 1 and 5

	Poverty in Year 5		
Poverty in Year 1	0	1	Total
0	516	234	750
1	211	190	401
Total	727	424	1151

Let's do a fixed effects logistic regression for the NLSY data, ignoring Years 2, 3, and 4, and focusing only on Years 1 and 5. Table 3.1 displays the cross-classification of Year 1 and Year 5. Although there was very little change in the marginal distribution of poverty across the two years, there were 234 girls whose family moved into poverty and 211 girls whose family moved out of poverty.

To do a fixed effects logistic regression, we first exclude the 706 girls who did not change over the 5 years. That's because a fixed effects analysis uses only within-person variation and, for these girls, there is no within-person variation on the response variable. So we are left with the 445 girls who experienced a change in poverty status. Within this reduced sample, let p_i be the probability that POV5 = 1, which is equivalent to the probability that a girl changed from 0 to 1 rather than from 1 to 0. We then apply conventional maximum likelihood to estimate the model

$$\log\left(\frac{p_i}{1 - p_i}\right) = (\mu_2 - \mu_1) + \beta(\mathbf{x}_{i2} - \mathbf{x}_{i1}) \tag{3.2}$$

That is, we do a logistic regression with POV5 as the dependent variable and difference scores for the time-varying predictors as the independent variables. This is actually a form of conditional maximum likelihood estimation, as explained in the next section. As in the linear case, both $\mathbf{z}_i$ and α_i drop out of the equation.[1]

Table 3.2 displays the estimates for three regression models, estimated with the **logit** command in Stata. Model 1 includes only the difference scores for the time-varying predictors. We see that motherhood increases the risk of poverty, while living with a spouse and working more hours reduces the risk. Again, keep in mind that these estimates control for all

Table 3.2 Logistic Regression on Difference Scores in Two-Period Case

	Model 1		Model 2		Model 3	
	Coefficient	Standard Error	Coefficient	Standard Error	Coefficient	Standard Error
Constant	0.539**	0.162	4.899**	1.644	3.052	1.826
DMOTHER	0.730**	0.250	0.744**	0.254	0.909**	0.270
DSPOUSE	−1.002**	0.283	−1.032**	0.292	−1.022**	0.301
DSCHOOL	0.343	0.212	0.339	0.218	0.639*	0.251
DHOURS	−0.0339**	0.0061	−0.0339**	0.0062	−0.0339**	0.0068
BLACK			−0.526*	0.216	−0.662**	0.226
AGE			−0.258*	0.103	−0.196	0.111
MOTHER1					0.457	0.460
SPOUSE1					0.442	0.726
SCHOOL1					1.184*	0.471
HOURS1					−0.0024	0.0128

* $.01 < p < .05$.

** $p < .01$.

time-invariant variables. Exponentiating the coefficient for motherhood (0.730), we get an odds ratio of 2.08. This tells us that when a girl has her first child, her odds of being in poverty are doubled. The intercept of 0.539 can be interpreted as the change in the log-odds of poverty from Year 1 to Year 5 for a girl who does not change on any of the time-varying predictors. Exponentiating, we get an odds ratio of 1.71—that is, an increase of 71% from Year 1 to Year 5.

Model 2 incorporates two time-invariant variables, BLACK and AGE, both of which have significant negative effects. The coefficients of these variables should be interpreted as interactions with time. Thus, for both variables, their effects on the risk of poverty were smaller (in magnitude) in Year 5 than in Year 1. Alternatively, the coefficients can be interpreted as showing how the rate of change over time varies across different subgroups. More specifically, for a girl whose time-varying predictors did not change from Year 1 to Year 5, the change in the log-odds of poverty over the 5-year period can be expressed as

$$4.899 - .526 \times \text{BLACK} - 2.58 \times \text{AGE}$$

Thus, for a 14-year-old girl who was not black and who did not change on any of the other predictors, the predicted change in the log-odds of being in poverty is 1.29. Equivalently, her odds of being in poverty should increase by a factor of $\exp(1.29) = 3.63$. On the other hand, blacks and girls who were older in Year 1 had a lower rate of increase in poverty.

Model 3 adds in the time-varying predictors measured in Year 1. As in Chapter 2, the coefficients of these variables can be interpreted as the *change* in the effect of each variable from Year 1 to Year 5, that is, interactions with time. Only one of these variables, SCHOOL1, is statistically significant. We can say, then, that the effect of enrollment in school on the log-odds of poverty increased by 1.184 from Year 1 to Year 5. The coefficient of 0.639 for DSCHOOL is the estimated effect of school enrollment in Year 5. So it looks like school enrollment had a negative effect in Year 1 (0.639 − 1.184) and a positive effect in Year 5.

In sum, the two-period method for logistic regression is very much like the two-period method for linear regression in Chapter 2. The big difference is that the logistic method requires that individuals who don't change on the response variable be excluded from the sample. For the dependent variable, I used the response variable at Period 2, which may seem somewhat different from the linear case. But if, instead, I had simply subtracted the Period 1 value from the Period 2 value, I would have gotten values of 1 and −1 rather than 1 and 0, which would amount to the same thing.[2]

Three or More Periods

How can this method be extended to incorporate information from all 5 years instead of just the first and fifth years? In Chapter 2, we accomplished this for linear models by creating a separate record for each time period in which each person was observed, pooling those records into a single data set, and then estimating a linear regression with dummy variables for persons. An alternative method also used multiple records for each person but avoided the dummy variables by expressing each variable as a deviation from its person-specific mean. Although these two methods produced identical results, the first method is properly described as unconditional maximum likelihood, while the second is a conditional maximum likelihood method.

Both conditional and unconditional maximum likelihood are available for logistic regression of dichotomous outcomes, but in this case they do not produce the same results. As in the linear case, unconditional maximum likelihood is implemented by creating multiple records per person and estimating a conventional logistic regression model with dummy variables for persons. Unfortunately, this method produces biased estimates of the coefficients (Hsiao, 1986). In fact, in the two-period case, the coefficient estimates are exactly twice as large as they should be (Abrevaya, 1997; Hsiao, 1986). The reason for this bias is something called the *incidental parameters problem* (Kalbfleisch & Sprott, 1970; Lancaster, 2000). What happens is that the number of parameters (in particular, the coefficients of the dummy variables for persons) increases directly with the sample size, thus violating one of the conditions that underlie the asymptotic theory of maximum likelihood estimation.

The solution is to do conditional maximum likelihood, which *conditions* the α_i parameters out of the likelihood function (Chamberlain, 1980). This is accomplished by conditioning the likelihood function on the total number of events observed for each person. In effect, each person's contribution to the likelihood function is the answer to a question such as the following: Given that a girl was in poverty for 2 out of the 5 years, what is the probability that this happened in, say, Years 2 and 4 (when it actually occurred) rather than in one of the nine other possible pairs of years? These conditional probabilities do not contain the α_i parameters. This conditioning approach only works for the logistic regression model for dichotomous response variables, not for other "link" functions such as probit or complementary log-log.

Many statistical packages have routines that will maximize this conditional likelihood for logistic regression. In Stata, this can be accomplished with either the **xtlogit** command or the **clogit** command.

The data structure required for these commands is similar to that described for the multiperiod case with **xtreg** in Chapter 2. There is one record for each period in which each individual is observed, with a common ID number for all records coming from the same individual. Values of the time-invariant predictors are replicated across the multiple records for each individual. If this method is applied to the two-period case, it produces exactly the same results as the difference method described above. For the 5-year example, Table 3.3 displays the first 15 observations in the working data set. These observations come from three girls, each observed for 5 years.

The Stata command **xtlogit** fits logistic regression models to panel data using three quite different methods: fixed effects (conditional likelihood), random effects, and generalized estimating equations. Table 3.4 presents results from applying all three methods to the teen poverty data. The first two columns show the results from conditional likelihood estimation of the fixed effects logit model. As with the two-period data, we see that motherhood and school enrollment are associated with increased risk of poverty, while living with a spouse and working more hours are associated with decreased risk.

How should we interpret these effects? Consider the coefficient of −0.748 for SPOUSE. Exponentiating, we get an odds ratio of 0.47. This says that if a girl changes from not living with a husband to living with a husband, her odds of being in poverty are multiplied by 0.47. In effect, getting married cuts her odds in half. For the HOURS coefficient of −0.0196, exponentiating yields an odds ratio of 0.98. This says that each additional hour of employment per week is associated with a 2% reduction in the odds of poverty. The YEAR coefficients are all comparisons with Year 1, and all are positive and statistically significant. Note that no constant is reported because the constant is conditioned out of the likelihood function.

The next two columns show results from estimating a logit model by the method of generalized estimating equations (GEE), which adjusts for dependence in the observations using iterated generalized least squares. Although the pattern is similar to the conditional likelihood analysis, there are three important differences. First, the SCHOOL coefficient has changed from positive and statistically significant to negative and not significant. Second, the coefficients for MOTHER and SPOUSE in the GEE analysis are noticeably larger in magnitude, while the year coefficients are smaller in magnitude. Third, the standard errors are all lower.

The last two columns of Table 3.4 give results from maximum likelihood estimation of a random effects model. This model is also represented by Equation 3.1 except that α_i is now assumed to be a set of random variables, each normally distributed with a mean of 0, a constant variance, and (most

Table 3.3 First 15 Observations in Teen Poverty Data Set

Observation	ID	YEAR	POV	MOTHER	SPOUSE	SCHOOL	HOURS	BLACK	AGE
1	22	1	1	0	0	1	21	0	16.00
2	22	2	0	0	0	1	15	0	16.00
3	22	3	0	0	0	1	3	0	16.00
4	22	4	0	0	0	1	0	0	16.00
5	22	5	0	0	0	1	0	0	16.00
6	75	1	0	0	0	1	8	0	17.00
7	75	2	0	0	0	1	0	0	17.00
8	75	3	0	0	0	1	0	0	17.00
9	75	4	0	0	0	1	4	0	17.00
10	75	5	1	0	0	1	0	0	17.00
11	92	1	0	0	0	1	30	0	16.00
12	92	2	0	0	0	1	27	0	16.00
13	92	3	0	0	0	1	24	0	16.00
14	92	4	1	1	0	0	31	0	16.00
15	92	5	1	1	0	0	0	0	16.00

Table 3.4 Conditional Likelihood and Other Estimates of the Logit Model

	Conditional Likelihood		GEE[a]		Random Effects	
	Coefficient	Standard Error	Coefficient	Standard Error	Coefficient	Standard Error
MOTHER	0.582**	0.160	0.850**	0.092	1.077**	0.119
SPOUSE	-.748**	0.175	-0.930**	0.121	-1.238**	0.152
SCHOOL	0.272*	0.113	-0.045	0.077	-0.064	0.098
HOURS	-0.0196**	0.0032	-0.0209**	0.0023	-0.0267**	0.0029
YEAR2	0.332**	0.102	0.223**	0.073	0.287**	0.100
YEAR3	0.335**	0.108	0.171*	0.080	0.226*	0.104
YEAR4	0.433**	0.116	0.196*	0.084	0.256*	0.108
YEAR5	0.402**	0.127	0.122	0.093	0.172	0.115
Constant			-0.543**	0.097	-0.681**	0.126

a. Specifying an unstructured correlation matrix and model-based standard errors.

*.01 < p < .05.

**p < .01.

important) independent of x_{it}. The random effects coefficient estimates are similar to those for GEE. The standard errors are all a bit larger than those for GEE, but smaller than those for the conditional likelihood analysis because conditional likelihood does not use any between-person variation. In fact, the conditional likelihood automatically excludes 324 girls (28%) whose poverty status did not change at all over the 5-year period. As in the two-period case, if people have no over-time variation, there is nothing for the predictors to explain.

Which of the three sets of results in Table 3.4 is the best? The key difference is that both GEE and random effects estimates do nothing to control for unmeasured predictors. In contrast, fixed effects estimation (conditional likelihood) controls for all constant predictors, with each girl acting as her own control. It also produces appropriate estimates of standard errors that are corrected for dependence. On the down side, those standard errors are larger than the random effects or GEE standard errors because a substantial amount of information in the data is not used. On balance, I prefer the fixed effects estimates for this application because they are much less vulnerable to omitted-variable bias. But in applications where the within-person variation is small relative to the between-person variation, the standard errors of the fixed effects coefficients may be too large to tolerate.

Another point worth noting is that both conditional likelihood and random effects estimates are "subject specific" while the GEE estimates are "population averaged." What's the difference? A subject-specific coefficient tells us what would happen to a single individual if that person's predictor variable were increased by one unit. In contrast, a population-averaged coefficient tells us what would happen to the whole population if everyone's predictor variable were increased by one unit. If the model is linear, there is no distinction between the two kinds of coefficients. For logistic regression models, however, as well as for many other nonlinear models, subject-specific coefficients are typically larger than population-averaged coefficients.

Which is better? The answer depends on your objectives. If you are a doctor and you want an estimate of how much a statin drug will lower your patient's odds of getting a heart attack, the subject-specific coefficient is the clear choice. On the other hand, if you are a state health official and you want to know how the number of people who die of heart attacks would change if everyone in the at-risk population took the statin drug, you would probably want to use population-averaged coefficients.

Even in the public health case, it could be argued that the subject-specific coefficient is more fundamental. Suppose that the true model is a basic random effects logistic model as expressed in Equation 3.1. The coefficient vectors β and γ are both subject-specific. But if we estimate the model with

GEE using **xtlogit**, we will get population-averaged coefficients $\beta*$ and $\gamma*$. The degree to which these coefficients differ depends on the variance of α_i. Specifically, if var$(\alpha_i) = 0$, then $\beta = \beta*$ and $\gamma = \gamma*$. As the variance of α_i increases, the values of $\beta*$ and $\gamma*$ will decline toward 0. When α_i has a normal distribution, the approximate relationship is

$$\beta * \approx \frac{\beta}{\sqrt{0.346 \, \text{var}(\alpha_i) + 1}} \tag{3.3}$$

So the population-averaged coefficients depend on the degree of unobserved heterogeneity in the logistic regression model. For the teen poverty data, the estimated variance of α_i is 1.454. Comparing the GEE and random effects coefficient estimates, we find that this relationship does, in fact, hold approximately.

Interactions With Time

Another downside of the conditional likelihood method is that coefficients can't be estimated for variables that don't vary over time (although these variables are implicitly controlled). Nevertheless, interactions between time-varying and time-constant variables can be included in the model. In Table 3.5, Model 1 includes a variable that is the product of MOTHER and BLACK, with a coefficient that is significant at the .05 level. Note that, unlike most models with interactions, it is not necessary (in fact, it's not even possible) to include the main effect of BLACK. The interaction can be interpreted just like interactions in a linear model. The coefficient of 0.982 for MOTHER represents the effect of MOTHER when BLACK = 0, that is, among girls who are not black. Exponentiating gives an odds ratio of 2.67. Thus, for nonblack girls, motherhood multiplies the odds of poverty by 2.67. To get the effect of motherhood among black girls, we add the main effect to the interaction coefficient, $0.982 - 0.599 = 0.383$, yielding a much lower odds ratio of 1.46.

In Model 2, we see significant interactions between YEAR and two time-varying predictors (SCHOOL and HOURS) and between YEAR and two time-constant predictors (BLACK and AGE). In this model, YEAR is treated as quantitative rather than categorical so as to simplify the model and its interpretation.[3] YEAR is coded with values of 0 through 4 (instead of 1 through 5) so that the main effects of SCHOOL and HOURS can be interpreted as the effects of these variables when YEAR = 0, that is, for the first year of observation. Also in the product terms, HOURS and AGE are expressed as deviations from their respective means to facilitate interpretation of the main effect of YEAR.

Table 3.5 Conditional Likelihood Estimates With Interactions

	Model 1		Model 2	
	Coefficient	Standard Error	Coefficient	Standard Error
MOTHER	0.982**	0.253	0.687**	0.163
SPOUSE	−0.783**	0.178	−0.741**	0.178
SCHOOL	0.267*	0.113	−0.311	0.190
HOURS	−0.0192**	0.0032	−0.0060	0.0063
YEAR2	0.332**	0.102		
YEAR3	0.334**	0.108		
YEAR4	0.430**	0.117		
YEAR5	0.400**	0.128		
MOTHER * BLACK	−0.599*	0.290		
YEAR			0.021	0.059
YEAR * SCHOOL			0.251**	0.063
YEAR * HOURS			−0.0055*	0.0021
YEAR * BLACK			−0.181**	0.048
YEAR * AGE			−0.056*	0.023

*$.01 < p < .05$.

**$p < .01$.

The interpretation of these interactions is somewhat different for the time-varying and time-constant predictors. For the time-varying predictors, it's usually best to consider how the effect of each predictor varies with time. For example, the effect of SCHOOL can be expressed as a linear function: $-0.311 + 0.251 \times$ YEAR. Thus, in the first year, the effect is negative and not statistically significant. Each additional year increases the effect by 0.251, so that by the fifth year the effect is 0.693 (or an odds ratio of 2). For HOURS, we have an effect of $-0.0060 - 0.0055 \times$ YEAR. Thus, the effect of HOURS starts out negative and gets steadily more negative each year, reaching -0.028 in the fifth year. This is equivalent to a 2.8% reduction in the odds of poverty for each additional hour worked.

For time-constant predictors, the best way to interpret the interactions is to examine how the effect of YEAR varies with these variables. Because of the coding of the variables, the main effect of YEAR (0.021) represents the estimated effect of YEAR among those who are not black, not enrolled in school, at the mean starting age of 15.65, and working the mean hours per week of 8.67. Among blacks (with these same characteristics), the effect of year is the main effect plus the interaction $(0.021 - 0.181 = -0.16)$. This corresponds to a 15% reduction in the odds of poverty for each additional year. We can also express the effect of YEAR as a linear function of AGE at the first year: $0.021 - .056 \times$ (AGE $- 15.65$). So for the 14-year-olds, the effect is 0.1134 (about a 12% increase in the odds of poverty per year) while for the 17-year-olds, the effect is -0.0546 (about a 5% reduction in the odds per year).[4]

A Hybrid Method

In Chapter 2, we combined the fixed effects and random effects approaches into a single model. This was accomplished by decomposing each time-varying predictor into a within-person component and a between-person component, and then fitting a random effects model with both components. The between-person component is just the person-specific mean of each variable. The within-person component is the deviation from that person-specific mean.

We now extend this method to logistic regression (Neuhaus & Kalbfleisch, 1998). As in the linear case, the attraction is that we can (a) include time-constant variables in the model, (b) perform tests for comparing fixed effects and random effects, and (c) fit a wider class of models. One example of (c) is that, unlike conditional likelihood, the hybrid approach can be used with other link functions such as probit or complementary log-log.

Again using Stata's **xtlogit** command, I fit a random effects model for the teen poverty data, with the results shown in Table 3.6. All the variable names beginning with M refer to person-specific means. All the variable names beginning with D refer to deviations from those means. The coefficients for the deviation variables are functionally equivalent to fixed effects coefficients because they are estimated using only within-person variation and therefore control for all stable predictors. For the linear hybrid model in Chapter 2, the deviation coefficients were identical to those produced by the least squares dummy variable method. In this case, the deviation coefficients are not identical to the conditional likelihood

Table 3.6 Hybrid Model for Teen Poverty Data

	Coefficient	*Standard Error*	*p*
DMOTHER	0.594	0.158	.000
DSPOUSE	−0.807	0.179	.000
DSCHOOL	0.275	0.113	.015
DHOURS	−0.0210	0.0032	.000
MMOTHER	1.079	0.181	.000
MSPOUSE	−2.146	0.255	.000
MSCHOOL	−1.362	0.202	.000
MHOURS	−0.0468	0.0058	.000
BLACK	0.572	0.097	.000
AGE	−0.123	0.050	.013
YEAR2	0.333	0.101	.001
YEAR3	0.330	0.107	.002
YEAR4	0.431	0.115	.000
YEAR5	0.391	0.125	.002
Constant	1.893	0.819	.021

Table 3.7   Tests of Equality for Deviation and Mean Coefficients

	Chi-Square	*p*
MOTHER	4.16	.041
SPOUSE	19.31	.000
SCHOOL	49.90	.000
HOURS	15.70	.000
Combined test (4 *df*)	79.10	.000

coefficients shown in Table 3.4, although they are certainly close. The standard errors are also very close to those in Table 3.4, leading to the same substantive conclusions.

The coefficients for the mean variables are not very interesting in themselves, but what's striking is how much larger (in magnitude) they are than the corresponding deviation coefficients. A conventional random effects model (one that doesn't split the within and between components) implicitly assumes that the deviation coefficients are identical to the mean coefficients. We can easily test that assumption within the hybrid model by directly testing for equality across the pairs of coefficients. The results in Table 3.7 clearly indicate a need to reject this assumption, suggesting that a fixed effects approach is superior to a random effects approach. In this table, the crucial test is the joint test that all four deviation coefficients are equal to the corresponding mean coefficients.[5]

Another advantage of the hybrid approach is the ability to get estimates for the time-constant predictors. We see in Table 3.6, for example, that blacks have significantly higher rates of poverty, while girls who were older at the first interview have significantly lower rates of poverty. It's important to keep in mind, however, unlike the coefficients for the deviation variables, that the coefficients of BLACK and AGE do *not* control for unmeasured predictors.

In Chapter 2, we saw that the hybrid linear model could be extended to allow for random coefficients for the time-varying predictors. That's also possible for the hybrid logit model, although estimating such models can be very computationally intensive. In Stata, logit models with random coefficients require a different command, **xtmelogit** (introduced in Stata 10). For the teen poverty example, I estimated a hybrid model that allowed a random coefficient for DMOTHER. The estimated mean for the coefficient was 0.603 with a standard deviation (*not* a standard error) of 0.751. The 95% confidence interval for this standard deviation was 0.272 to 2.075. Since this confidence interval does not include zero, there is some evidence for variability across persons in the effect of motherhood on poverty.

Methods for More Than
Two Categories on the Response Variable

To this point, we have only considered dichotomous response variables. Now consider a categorical response variable y_{it} that can take on more than two values. Suppose that those values are the integers ranging from 1 to J, with the running index j. Let $p_{itj} = \text{Prob}(y_{it} = j)$. We then need a model for how this probability depends on our predictor variables $\mathbf{x}_{it}$ and $\mathbf{z}_i$.

Let's begin with the case in which the categories of the dependent variable are ordered. The most popular model for ordered categorical response variables is the *cumulative logit model*, also known simply as the ordered logit model. A fixed effects version of the model can be written as

$$\log\left(\frac{F_{ijt}}{1 - F_{ijt}}\right) = \mu_{tj} + \beta \mathbf{x}_{it} + \gamma \mathbf{z}_i + \alpha_i, \quad j = 1, \ldots, J-1 \quad (3.4)$$

where $F_{ijt} = \sum_{m=j}^{J} p_{imt}$ is the "cumulative" probability of being in category j or higher. Unfortunately, conditional maximum likelihood is not possible for this model because it does not have "reduced sufficient statistics" for the α_i parameters. What we can do, however, is apply the hybrid method discussed in the last section, using conventional maximum likelihood with robust standard errors to adjust for lack of independence in the repeated observations for each individual.

As an example, let's return to the antisocial behavior example of Chapter 2. In that data set, the dependent variable ANTI had integer values of 0 through 6 but was treated as a quantitative dependent variable in a linear regression model. Here we take the arguably more appropriate approach of treating ANTI as an ordered categorical variable in a logistic regression model.

As in the binary case, the hybrid method is implemented by calculating person-specific means for each of the predictor variables and then calculating deviations from those means. Both mean variables and deviation variables are included as predictors in the cumulative logit model. To get subject-specific coefficients, we would need to estimate a random effects model, but it's not easy to find commercial software that will do that for an ordered logit model. Instead, we will just do conventional maximum likelihood estimation with robust standard errors that correct for dependence in the repeated observations. I did this in Stata using the `ologit` command, yielding the results shown in Table 3.8. These results are remarkably similar to those for the hybrid linear model in Table 2.8. The p values would certainly lead to the same conclusion for every coefficient. Even the coefficients and standard errors are quite close, although that should not be expected in most applications. The coefficients for the cumulative logit model represent the change in the log-odds of being in a higher rather than a lower category of the dependent variable. As in a binary regression model, exponentiating these coefficients produces odds ratios. For more details on how to interpret these coefficients, see Allison (1999a).

Table 3.8 Hybrid Cumulative Logit Model for Antisocial Behavior

	Coefficient	Robust SE	p
DSELF	−0.064	0.013	.000
DPOV	0.116	0.117	.320
MSELF	−0.108	0.027	.000
MPOV	0.696	0.185	.000
BLACK	0.153	0.157	.330
HISPANIC	−0.310	0.169	.065
CHILDAGE	0.083	0.111	.453
MARRIED	−0.189	0.163	.247
GENDER	−0.598	0.128	.000
MOMAGE	−0.017	0.029	.557
MOMWORK	0.190	0.146	.195
TIME_2	0.016	0.069	.819
TIME_3	0.167	0.077	.030

The coefficients for the two deviation variables, DSELF and DPOV, can be interpreted as if they were fixed effects coefficients. Because these coefficients depend only on variation over time within persons, they control for all stable predictors. A test of the null hypothesis that the two deviation coefficients are equal to the two corresponding mean coefficients (using the **test** command in Stata) yields a chi-square of 9.02 with 2 *df*, which is significant at the .01 level. As in the linear case, most of the action comes from the difference in the coefficients for MPOV and DPOV. The implication is that a conventional ordered logit model, even with standard errors corrected for dependence, would not be appropriate here, at least not for the POV variable. Instead, we should focus our attention on the deviation coefficients since they control for all stable predictors.

We now turn to the more complicated situation in which the categories of the dependent variable are not ordered. The most widely used model for unordered categorical variables is the multinomial logit model, also known as the generalized logit model. Here is a fixed effects version of that model:

$$\log\left(\frac{p_{ij}}{p_{iJ}}\right) = \mu_{tj} + \beta_j \mathbf{x}_{it} + \gamma_j \mathbf{z}_i + \alpha_{ij}, \quad j = 1, \ldots, J - 1 \tag{3.5}$$

Equation 3.4 can be viewed as a set of simultaneous binary logistic regression equations, each equation comparing one category with last category. The fixed effects α_{ij} vary both over individuals and over possible response values, but not over time.

The fixed effects multinomial logit model is like the binary logit model in that it has reduced sufficient statistics for the α_{ij}s, namely, the frequency counts of the different response values for each individual. In principle, the model can be estimated by conditional maximum likelihood with conditioning on those counts (Chamberlain, 1980). However, there is no commercial software available to do this. If the time-varying predictors are categorical, the model can be reformulated as a log-linear model and estimated in that framework (Conaway, 1989; Darroch & McCloud; 1986, Kenward & Jones, 1991; Tjur, 1982). Setting up the model in that form is somewhat complicated, however, so I will not consider that option here.

Another approach to estimation is to decompose the multinomial model into a set of binary models, one model for each comparison of a particular category with a reference category (Allison, 1999a; Begg & Gray, 1984). Each binary model can then be estimated using the conditional logistic regression methods that we have already discussed in this chapter. While this approach produces approximately unbiased estimates of the coefficients, results will differ depending on the choice of the reference category. Furthermore, there is no overall test for the effect of each variable on the response variable.

As we saw with the cumulative logit model, the hybrid method is the most easily implemented approach to estimating the multinomial logit model with fixed effects. As an example, let us return to the sample used for most of this chapter, which has 1,151 teenaged girls interviewed annually for 5 years. Now, however, we shall use a new response variable EMPSTAT, which has three categories:

1. if currently employed,

2. if unemployed (laid off or seeking work), and

3. if out of the labor force (in school, keeping house, etc.).

As predictor variables, we shall use MOTHER (currently has at least one child), SPOUSE (currently lives with husband), current AGE, and BLACK (vs. nonblack). The first three of these variables are time varying.

The first step is to calculate the means of the time-varying variables for each individual and the deviations from those means. Because there were 241 records with missing data on the response variable EMPSTAT, it's important to calculate these means *after* deleting the records with missing data.

To estimate the multinomial logit model in Stata, I used the **mlogit** command with robust standard errors to correct for dependence in the repeated observations for each person.

Results are shown in the first two columns of numbers in Table 3.9. There we see binary logistic regression equations, each comparing one category of EMPSTAT with Category 1 (employed) as the reference category. While the coefficients are conventional maximum likelihood estimates under the assumption that all observations are independent, the standard errors are adjusted for dependence. Keep in mind that these are population-averaged rather than subject-specific coefficients.

Focusing on the deviation coefficients, we see that motherhood increases the odds of being unemployed or out of the labor force. Living with a husband increases the odds of being unemployed (vs. employed) but reduces the odds of being out of the labor force. As girls get older, they are slightly less likely to be unemployed and considerably less likely to be out of the labor force. I also performed tests of whether the coefficients for each deviation variable are the same as the coefficients for the corresponding mean variable, thus testing the fixed effects model against a conventional logistic regression model. For both equations, the chi-square tests were highly significant.

Table 3.9 Hybrid Multinomial Logit Model for Employment Status

| | GEE (Population Averaged) | | | | Random Effects (Subject Specific) | | | |
| | Unemployed vs. Employed | | Out of the Labor Force vs. Employed | | Unemployed vs. Employed | | Out of the Labor Force vs. Employed | |
	Coefficient	Robust SE	Coefficient	Robust SE	Coefficient	SE	Coefficient	SE
D MOTHER	0.927**	0.160	0.799**	0.155	1.237**	0.201	0.951**	0.221
M MOTHER	1.656**	0.163	0.389*	0.167	2.185**	0.212	0.611**	0.234
D SPOUSE	0.640**	0.180	0.569*	0.221	0.816**	0.211	0.617*	0.269
M SPOUSE	0.678**	0.242	-1.111**	0.264	0.829**	0.305	-1.459**	0.394
D AGE	-0.070*	0.029	-0.381**	0.027	-0.131**	0.037	-0.503**	0.033
M AGE	-0.306**	0.045	-0.505**	0.046	-0.394**	0.061	-0.663**	0.062
BLACK	0.393**	0.096	0.499**	0.093	0.597**	0.127	0.720**	0.130
Constant	4.381**	0.799	8.612**	0.802	5.630**	1.076	11.29**	1.091

*.01 < p < .05.

**p < .01.

In the last two columns of Table 3.9, we see alternative estimates of the multinomial logit model obtained by estimating two separate random effects models with the **xtlogit** command in Stata. For the first model (unemployed vs. employed), all the records in the "out of the labor force" category were dropped. For the second model (out of the labor force vs. employed), all the records in the "unemployed" category were dropped. Comparing the random effects estimates with the GEE estimates, we find that signs and significance levels for all coefficients are about the same. However, the random effects estimates are generally somewhat larger in magnitude because they are subject-specific rather than population averaged.

Summary

All the fixed effects methods described in Chapter 2 for linear models can be extended to categorical response variables. The basic message is the same. Fixed effects methods control for the potential confounding effects of all unobserved, time-invariant variables. On the other hand, fixed effects methods tend to be less efficient than other methods because between-individual variation is ignored. For categorical outcomes, somewhat different estimation procedures are required.

The main focus of the chapter was on regression models for dichotomous responses. When each individual has exactly two dichotomous observations, a fixed effects logistic model can be estimated by conditional maximum likelihood using a conventional logistic regression program. The method is to (a) discard all cases that have the same values on the two response variables, (b) recode all time-varying predictors as difference scores, and (c) fit an ordinary binary logistic regression to one of the response variables.

When individuals have more than two dichotomous response observations, a different data structure is needed—namely, a separate record for each response for each individual. However, because of the "incidental parameters problem," one cannot simply estimate a conventional logistic regression model with dummy variables for the individuals. The coefficients from such a regression will be biased away from zero, especially if the number of observations per individual is small. The solution is to use conditional maximum likelihood to remove the fixed effects from the likelihood function. In Stata, this is accomplished with the **xtlogit** or **clogit** commands.

Instead of fixed effects logistic regression, many researchers use GEE estimation or random effects logistic regression, both of which can be implemented with the **xtlogit** command. In contrast to the fixed effects approach, however, neither of these methods controls for unmeasured, time-invariant explanatory variables. Like fixed effects, random effects estimation produces "subject-specific" rather than "population averaged" coefficients. The latter are generally attenuated toward zero by unobserved heterogeneity.

The fixed effects and random effects approaches may be combined into a hybrid model by estimating a random effects model in which the time-varying predictors are decomposed into individual means and deviations from those means, and then estimating a random effects model. As we saw in Chapter 2, the hybrid method allows one to include time-invariant variables in the model and provides a simple test comparing the fixed effects model with the random effects model.

For response variables with more than two categories, conditional maximum likelihood estimation for fixed effects logistic regression is generally unavailable in commercial software. Instead, the best available approach at present for both ordinal and nominal response variables is to use the hybrid method with robust standard errors to correct for dependence.

Notes

1. For a derivation of this result, see Allison (2005).

2. In the two-period case, conditional likelihood can also be implemented using the method described in the next section for three or more periods, using the **xtlogit** command in Stata. Results will be identical to the "difference" method just described.

3. I also fit a model with YEAR treated as categorical in all four interactions, but this model did not differ significantly from Model 2 by a likelihood ratio test.

4. In Table 3.5, I only examined selected interactions. In many applications, it may be desirable to do a comprehensive test of the stability of the model over time by simultaneously testing all possible interactions with time. This can be accomplished by comparing two models: the model containing all interactions with time and the model that excludes those interactions. The difference in the likelihood ratio chi-square statistics for the two models is itself a likelihood ratio chi-square test of the null hypothesis that all the interaction coefficients are zero (with degrees of freedom equal to the difference in the *df* for the two models).

5. The tests are easily accomplished with the **test** command in Stata. See Appendix 1 for details.

CHAPTER 4. FIXED EFFECTS
MODELS FOR COUNT DATA

Often our dependent variables are counts of something: number of children, number of sex partners in the last year, number of computers in the home, number of arrests in the past 5 years, and so on. Many researchers treat count variables as continuous measures and do their analysis with ordinary least squares regression. But this may be inappropriate for several reasons. Count variables are necessarily discrete and cannot have values less than zero. Usually, their distributions are highly skewed.

A generally superior approach is to estimate a Poisson regression model or a negative binomial regression model, both of which are explicitly designed to model count data (Long, 1997). After a brief review of these methods, we will see how to extend these count data methods to handle multiple periods per individual along with fixed effects to control for all time-invariant predictor variables.[1] In the process, we will revisit many of the issues that arose for dichotomous outcomes in the previous chapter. However, the estimation problems that plague logistic models turn out to be less serious for count data models.

We begin by considering the example that will be used throughout the chapter. The data consist of 346 manufacturing firms with yearly counts of patents received in each of the years from 1975 to 1979. Previous analyses of these data can be found in Hall, Griliches, and Hausman (1986) and Cameron and Trivedi (1998). In the original data set, there is one record per firm, with variables PAT75 to PAT79 containing the patent counts in the 5 years. As predictors we have the logarithm of research and development (R&D) expenditures in each year between 1970 and 1979 (LOGR70 to LOGR79). There are also two time-invariant predictors: LOGSIZE, which is the logarithm of the book value of the firm in 1972, and SCIENCE, a dummy variable equal to 1 if the firm is in the science sector, otherwise 0.

Poisson Models for Count Data
With Two Periods per Individual

As we have seen in the previous chapters, when there are only two periods per individual, a linear or logistic fixed effects analysis can be done using simplified methods with conventional software. This is also true for count data. In fact, a fixed effects Poisson regression model can be estimated with an ordinary logistic regression program for grouped data.

To illustrate this for the patent data, we will ignore the intervening years and focus only on 1975 and 1979. Let y_{i1} be the patent count for firm i in 1975 and y_{i2} the patent count in 1979. Each of these variables is assumed to have a Poisson distribution with an expected value of λ_{it}. That is, the probability that $y_{it} = r$ is given by

$$\Pr(y_{it} = r) = \frac{\lambda_{it}^{r} e^{-\lambda_{it}}}{r!}, \quad r = 0, 1, 2, \ldots \tag{4.1}$$

The Poisson distribution is perhaps the simplest probability distribution that is appropriate for count data. It may be derived from a stochastic process model under the assumptions that (a) events (in this case patents) cannot occur simultaneously and (b) events are independent (Cameron & Trivedi, 1998). The independence assumption means that the occurrence of an event neither raises nor lowers the probability of future events.

Note that our model does not assume that there is a single Poisson distribution for the entire sample. Instead, each firm's patent count is drawn from a different Poisson distribution whose expected value λ_{it} varies across firms and over time. An unusual property of the Poisson distribution is that its mean and variance are equal:

$$E(y_{it}) = \text{var}(y_{it}) = \lambda_{it} \tag{4.2}$$

Unfortunately, as we shall see, this property sometimes leads to a problem called overdispersion that can seriously compromise the estimation of Poisson regression models.

Next, we let λ_{it} be a log-linear function of the predictor variables

$$\log \lambda_{it} = \mu_t + \beta \mathbf{x}_{it} + \gamma \mathbf{z}_i + \alpha_i \tag{4.3}$$

As in earlier chapters, $\mathbf{x}_{it}$ represents the time-varying predictor variables, $\mathbf{z}_i$ denotes the time-invariant predictors, and α_i denotes the unobserved "fixed effects." As before, treating α_i as a set of fixed constants is equivalent to treating them as random variables that have unrestricted correlations with $\mathbf{x}_{it}$. The vector $\mathbf{x}_{it}$ includes the R&D expenditures in the current year t and in each of the preceding 5 years.

Our objective is to estimate the parameters in Equation 4.3. To do this, we shall use conditional maximum likelihood, the same method used in Chapter 3 to estimate the fixed effects logistic regression model. Consider the distribution of y_{i2} conditional on the total event count for the two time periods combined, denoted by $w_i = y_{i1} + y_{i2}$. It can be shown that

$y_{i2}|w_i \sim B(p_i, w_i)$. That is, conditional on the total count w_i, the 1979 count y_{i2} has a binomial distribution with parameters p_i and w_i, where

$$p_i = \frac{\lambda_{i2}}{\lambda_{i2} + \lambda_{i1}} \tag{4.4}$$

It follows, after a bit of algebra, that

$$\log\left(\frac{p_i}{1 - p_i}\right) = (\mu_2 - \mu_1) + \beta(\mathbf{x}_{i2} - \mathbf{x}_{i1}) \tag{4.5}$$

Thus, we have converted our Poisson regression model into a logistic regression model in which the predictor variables are difference scores for the original predictors. Note that, as in earlier applications, both α_i and γz_i drop out of Equation 4.5.

To implement this conditional method in Stata, I used the **blogit** command, which does ML estimation of grouped binomial data. The **blogit** command expects the dependent variable to come in two parts: the number of "events" and the number of "trials." I first fit a model with no predictors (just the constant term) by submitting the command

blogit pat79 total

where PAT79 is the number of patents in 1979 and TOTAL is equal to PAT75 + PAT79. The estimated constant was −0.1386 with a standard error of 0.0129, yielding a z statistic of −10.68. What does this tell us? Well, if m_1 is the mean number of patents in Year 1 and m_2 is the mean for Year 2, the constant is just $\log(m_1/m_2)$. If the mean number of patents were exactly the same in both years, the constant would be 0. The fact that it's negative indicates that the mean went down over time. More specifically, if we calculate

$$100(\exp(-0.1386) - 1) = -12.9\%$$

we get the percentage decrease in the mean from 1975 to 1979. Furthermore, because the z statistic for the constant is so large, we can reject the null hypothesis that the means for the two years are the same.

Actually, this z statistic is *too* large. Because of a problem called *overdispersion*, the conventional standard error estimate based on the Poisson distribution is an underestimate of the true standard error. I'll discuss overdispersion in more detail later in this chapter. In the meantime, we can get a better standard error estimate in Stata using either the jackknife or bootstrap options. (These computer-intensive methods estimate standard errors by replicating the analysis on many subsamples or resamples of the data. For a detailed explanation, see Mooney & Duval, 1993.) The jackknife standard error is 0.0371, producing a z statistic of -3.74. The bootstrap standard error is 0.0358, yielding a z statistic of -3.87. Although these z statistics are much smaller than the conventional z statistic, they are obviously still highly significant.

The next step is to introduce predictors that are difference scores for the logged measures of R&D expenditures. To be consistent with previous analyses of these data, the goal is to include both the "current" value of R&D and the lagged values for each of the previous 5 years. I defined the following variables:

$$RD0 = LOGR79 - LOGR75$$
$$RD1 = LOGR78 - LOGR74$$
$$RD2 = LOGR77 - LOGR73$$
$$RD3 = LOGR76 - LOGR72$$
$$RD4 = LOGR75 - LOGR71$$
$$RD5 = LOGR74 - LOGR70$$

RD0 is the difference score for the same years in which the patents were counted, while RD1 through RD5 are difference scores for lags of 1 to 5 years. All six of these variables were included as predictors in the grouped logistic regression model, with the results shown in Model 1 of Table 4.1.

Examining the parameter estimates and their associated statistics, we see that RD0, the contemporaneous measure of R&D expenditures, has a highly significant effect on the patent count, with a coefficient of 0.5214. To interpret this, keep in mind that both the dependent variable (expected number of patents) and the independent variable (R&D expenditures) are logged (see Equation 4.3). Because both variables are logged, we may say that a 1% increase in R&D expenditures is associated with a 0.52% increase in the expected number of patents in the same year, controlling for the lagged R&D measures. The effects of the lagged measures are much smaller.

Again, we need to deal with the overdispersion problem by using more robust estimates of the standard errors. The bootstrap standard errors shown in Table 4.1 are more than twice as large as the conventional standard errors. Using the bootstrap standard errors, we find that only RD0 retains its statistical significance, and even for this variable the z statistic is greatly reduced.

As with our earlier fixed effects models, the estimates in Table 4.1 control for all variables that are constant over time. We can put predictor variables that do *not* vary with time in the model, although the interpretation of their

Table 4.1 Conditional Poisson Estimates for Patent Data, Two Time Periods

	Model 1			Model 2		
	Coefficient	Conventional Standard Error	Bootstrap Standard Error	Coefficient	Conventional Standard Error	Bootstrap Standard Error
RD0	0.521	0.084**	0.207*	0.533	0.085**	0.209*
RD1	-0.207	0.113	0.227	-0.192	0.113	0.256
RD2	-0.118	0.111	0.277	-0.137	0.111	0.341
RD3	0.060	0.096	0.263	0.062	0.096	0.314
RD4	0.181	0.090*	0.244	0.183	0.091*	0.209
RD5	-0.093	0.069	0.188	-0.100	0.069	0.167
SCIENCE				0.023	0.028	0.089
LOGSIZE				0.017	0.008*	0.017
Constant	-0.222	0.018**	0.052**	-0.347	0.062**	0.138*

*.01 < p < .05.

**p < .01.

coefficients is not so straightforward. Model 2 of Table 4.1 includes the dummy variable for SCIENCE sector and the LOGSIZE of the firm. Neither variable approaches statistical significance when we use bootstrap standard errors. Their coefficients can be interpreted as measuring interactions between each variable and time. Like all interactions, these coefficients can be interpreted in two ways. For example, the coefficient of 0.0275 for SCIENCE represents the *difference* between the coefficient for SCIENCE in 1979 and the coefficient in 1975. The fact that it is far from statistically significant suggests that this variable has the same effect in both the years. Alternatively, we can interpret 0.0275 as the increment in the effect of time for firms in the science sector, relative to those not in the science sector. Again, because it is far from significant, we may conclude that the rate of change in the number of patents from 1975 to 1979 is essentially the same for firms in the two sectors. Similar interpretations can be made for LOGSIZE.

Poisson Models for Data With More Than Two Periods per Individual

When individuals are observed at more than two periods, estimation of a fixed effects Poisson model requires a different approach. We now continue the example of the last section by analyzing annual patent counts from 1975 to 1979, with each count denoted by y_{it}. As before, we assume that each y_{it} is drawn from a Poisson distribution given by Equation 4.1 with expected value λ_{it}, and we let λ_{it} be the log-linear function of the predictor variables given in Equation 4.3.

There are two approaches to estimation, conditional ML and unconditional ML. In conditional ML, the likelihood function is conditioned on the sum (over time) of all the counts for each individual, which eliminates the fixed effects (α_i). The resulting conditional likelihood (Cameron & Trivedi, 1998) is proportional to

$$\prod_i \prod_t \left(\frac{\exp(\mu_t + \beta\mathbf{x}_{it})}{\sum_s \exp(\mu_s + \beta\mathbf{x}_{is})} \right)^{y_{it}} \tag{4.6}$$

In Stata, this likelihood can be maximized with the **xtpoisson** command (which can also estimate random effects and population-averaged models). This command requires that the data be restructured so that there is one record for each firm year, with a common ID variable linking together the five records for each firm.[2] The new data set has 1,730 observations for 346 firms. Table 4.2 displays the 20 records for the first four firms in the sample.

Table 4.2 Observations for the First Four Firms in Restructured Data Set

Observation	ID	t	PATENT	RD0	RD1	RD2	RD3	RD4	RD5
1	1	1	32	0.92327	1.02901	1.06678	0.94196	0.88311	0.99684
2	1	2	41	1.02309	0.92327	1.02901	1.06678	0.94196	0.88311
3	1	3	60	0.97240	1.02309	0.92327	1.02901	1.06678	0.94196
4	1	4	57	1.09500	0.97240	1.02309	0.92327	1.02901	1.06678
5	1	5	77	1.07624	1.09500	0.97240	1.02309	0.92327	1.02901
6	2	1	3	-1.48519	-0.68464	-0.15087	0.08434	-0.21637	-0.45815
7	2	2	2	-1.19495	-1.48519	-0.68464	-0.15087	0.08434	-0.21637
8	2	3	1	-0.60968	-1.19495	-1.48519	-0.68464	-0.15087	0.08434
9	2	4	1	-0.58082	-0.60968	-1.19495	-1.48519	-0.68464	-0.15087
10	2	5	1	-0.60915	-0.58082	-0.60968	-1.19495	-1.48519	-0.68464

(Continued)

Table 4.2 (Continued)

Observation	ID	t	PATENT	RD0	RD1	RD2	RD3	RD4	RD5
11	3	1	49	3.67343	3.58542	3.52962	3.44199	3.40697	3.39054
12	3	2	42	3.77871	3.67343	3.58542	3.52962	3.44199	3.40697
13	3	3	63	3.82205	3.77871	3.67343	3.58542	3.52962	3.44199
14	3	4	77	3.88021	3.82205	3.77871	3.67343	3.58542	3.52962
15	3	5	80	3.90665	3.88021	3.82205	3.77871	3.67343	3.58542
16	4	1	0	0.43436	0.53714	0.48840	0.58779	0.48454	0.54340
17	4	2	0	0.33836	0.43436	0.53714	0.48840	0.58779	0.48454
18	4	3	1	0.36561	0.33836	0.43436	0.53714	0.48840	0.58779
19	4	4	0	0.43860	0.36561	0.33836	0.43436	0.53714	0.48840
20	4	5	0	0.42459	0.43860	0.36561	0.33836	0.43436	0.53714

As in the two-period case, our regression model includes R&D expenditure for the current year and in the five preceding years. The model also includes dummy variables corresponding to 4 out of the 5 years (the first year is the reference category). Results in Table 4.3 are similar to those we got in Table 4.1 using just 2 of the 5 years. That is, we find strong effects of R&D expenditures in the same year (RD0), but much weaker effects for the lagged values (RD1 to RD5). The TIME coefficients show a marked tendency for patent counts to decline over the 5-year period. Note that no constant is reported for the fixed effects regression because the constant drops out of the conditional likelihood function.

Standard error estimates for the fixed effects (conditional likelihood) were estimated both by the conventional method and the bootstrap method.[3] As in the two-period analysis, the bootstrap standard errors are much larger than the conventional standard errors, nearly twice as large in most cases. Again, the reason for this is overdispersion, a very common problem with Poisson regression. Basically, overdispersion means that there is more variation in the event counts than would be expected based on a Poisson distribution. This usually happens because the regression model does not include all the causes of variation in the counts. Because we are estimating a fixed effects model, however, we have already controlled for all the between-firm variability in the patent counts. So the only omitted variables that could produce overdispersion are those that vary over time within firm. With some software for Poisson regression (e.g., SAS), you get a statistic called the *deviance* that directly measures the degree of overdispersion. Stata doesn't report the deviance for conditional Poisson regression, however, so it's always a good idea to use the bootstrap or jackknife standard errors to avoid potential errors.

For comparison, Table 4.3 also reports results for the other two types of models estimated by **xtpoisson**, the random effects model and population-averaged model (estimated by generalized estimating equations, or GEE). Like the fixed effects model, the random effects model is also described by Equations 4.1 and 4.3 except that α_i is assumed to be a random variable that has a specified probability distribution and is *independent* of x_{it} and z_i. This independence assumption implies that the random effects model does *not* control for unobserved covariates.

The default in Stata is to assume that α_i has a log-gamma distribution, but it's also possible to specify a normal distribution. The population-averaged model, on the other hand, does not postulate an additional disturbance term in the Poisson regression equation, but merely allows the multiple observations for each firm to be correlated.[4] This model is estimated by the GEE method, which, as in the logistic case, is a kind of iterated generalized least squares. Both random effects and GEE are vulnerable to overdispersion, so the conventional

Table 4.3 Poisson Regression Estimates for Patent Data, Five Time Periods

	Fixed Effects			Random Effects		GEE	
	Coefficient	Conventional Standard Error	Bootstrap Standard Error	Coefficient	Bootstrap Standard Error	Coefficient	Robust Standard Error
RD0	0.322	0.046**	0.084**	0.477	0.072**	0.303	.053**
RD1	−0.087	0.049	0.087	−0.008	0.058	0.049	.056
RD2	0.079	0.045	0.064	0.136	0.061*	0.167	.051**
RD3	0.001	0.041	0.072	0.059	0.090	0.085	.062
RD4	−0.005	0.038	0.065	0.028	0.051	0.050	.042
RD5	0.003	0.032	0.063	0.082	0.067	0.038	.043
TIME 2	−0.043	0.013**	0.017*	−0.047	0.016**	−0.048	.017**
TIME 3	−0.040	0.013**	0.026	−0.056	0.024*	−0.052	.026*
TIME 4	−0.157	0.014**	0.036**	−0.190	0.041**	−0.178	.043**
TIME 5	−0.198	0.015**	0.033**	−0.253	0.038**	−0.234	.041**
Constant				1.403	0.081**	1.828	.123**

$*.01 < p < .05.$

$**p < .01.$

standard errors are biased. For random effects, I report bootstrap standard errors. For GEE, I report robust standard errors that are more easily calculated but are unavailable in **xtpoisson** for random effects and fixed effects.

As we saw in the previous comparisons, the fixed effects estimates tend to have larger standard errors than those for random effects and GEE models. As usual, that's because fixed effects only uses within-firm variation and essentially discards any between-firm variation. In fact, firms that have zero patents in all 5 years are completely eliminated from the conditional likelihood function. There were 22 such firms in this data set. On the positive side, the fixed effects estimates control for all constant firm characteristics, whereas the random effects and GEE estimates only control for those firm-level characteristics that are explicitly included in the models (and none are included in these models). For this analysis, the only major difference in the results across the three methods is that there is some evidence for an effect of RD2 in the random effects and GEE models but no such evidence with fixed effects.

The fixed effects Poisson regression model can also be estimated by *unconditional* ML. This is accomplished by estimating a conventional Poisson regression model that includes dummy variables for all the firms (less one). In Chapter 3 for the logistic regression model, we saw that conditional and unconditional ML produced different estimates. Furthermore, the unconditional estimates were wrong—they tended to produce coefficient estimates that were too large. For the Poisson regression model, on the other hand, conditional and unconditional estimation always produce *identical* results (Cameron & Trivedi, 1998). Consequently, the choice between one or the other is purely a matter of computational convenience. In Stata, the unconditional method takes *much* longer to compute for the patent data because coefficients for more than 300 dummy variables must be estimated. But many statistical packages (e.g., SAS) do not have procedures for conditional Poisson regression, in which case the unconditional method is the only option.

The predictor variables in Table 4.3 are all time varying. Can we also include time-invariant predictors in the fixed effects model? In the previous section, with only two observations per firm, we included two time-invariant predictors in the logistic regression model used for conditional estimation of the Poisson model. The coefficients of those variables were interpreted as interactions with time. In the present setup, time-invariant predictors cannot be directly included in the model. However, we can specify interactions between time-invariant predictors and time-varying predictors, including time itself. For example, one might hypothesize that R&D expenditures have a greater effect on patents in science-based firms than in other sectors. Table 4.4 reports results for a model that includes a

Table 4.4 Conditional Poisson Estimates With Time-Invariant Covariate

	Fixed Effects		
	Coefficient	Conventional Standard Error	Bootstrap Standard Error
RD0	0.375	0.048**	0.078**
RD0 * SCIENCE	−0.204	0.067**	0.188
TIME_2	−0.034	0.013**	0.014*
TIME_3	−0.034	0.013**	0.020
TIME_4	−0.151	0.014**	0.031**
TIME_5	−0.189	0.015**	0.035**

*.01 < p < .05.

**p < .01.

variable that is the product of SCIENCE and RD0. Note that it is not necessary (and not even possible) to include the main effect of SCIENCE. For simplicity, this model deletes the lagged effects of R&D, which were not significant in Table 4.3.

In Table 4.4, we see that with conventional standard errors there is a significant interaction between RD0 and the dummy variable SCIENCE, but with bootstrap standard errors the interaction is not significant. In any case, the interaction is opposite to the hypothesis—R&D expenditures have a greater impact on patent counts among nonscience firms than in science firms. To be more specific, the effect of R&D for nonscience firms is 0.375, the main effect for RD0. The effect for science firms is 0.375 − 0.204 = 0.171, the main effect plus the interaction.

Now let's test whether the rate of change over time in patents is different for science and nonscience sectors. For the model in Table 4.5, I constrained the effect of time to be linear and then included an interaction between SCIENCE and TIME. The results in Table 4.5 show no evidence for a difference between science and nonscience firms in their rate of change over time. The interaction coefficient is far from statistically significant (either by conventional or bootstrap standard errors), and its magnitude is only about 2% of the main effect of time.

Table 4.5 Conditional Poisson Estimates Including Interaction With Time

	Fixed Effects		
	Coefficient	Conventional Standard Error	Bootstrap Standard Error
RD0	0.276	0.039	0.075
TIME	−0.049	0.005	0.010
SCIENCE * TIME	−0.001	0.006	0.016

Fixed Effects Negative Binomial Models for Count Data

As we just saw, fixed effects Poisson regression models are quite vulnerable to the effects of overdispersion. That's somewhat unexpected because fixed effects already allow for unobserved heterogeneity across individuals by way of the α_i parameters. That heterogeneity is presumed to be time invariant, however, and there may still be unobserved heterogeneity that is specific to particular points in time, leading to observed overdispersion. As we have seen, the standard errors can be corrected for overdispersion by using the bootstrap or jackknife methods. Although that's not a bad approach, we can do better by directly building overdispersion into the model for event counts.

To model the overdispersion, we now assume that the patent counts are drawn from a negative binomial distribution for each firm at each point in time. The negative binomial distribution is a generalization of the Poisson distribution that allows for overdispersion by way of an additional parameter. The appeal of the negative binomial model is that the estimated regression coefficients may be more efficient (less sampling variability), and the standard errors and test statistics may be more accurate than those produced by such empirical, after-the-fact corrections, as the bootstrap or jackknife.

There is more than one way to formulate a negative binomial regression model, however. The model used here is what Cameron and Trivedi (1998) call an NB2 model, in which the probability mass function for y_{it} is given by

$$\Pr(y_{it} = r) = \frac{\Gamma(\theta + r)}{\Gamma(\theta)\Gamma(r + 1)} \left(\frac{\lambda_{it}}{\lambda_{it} + \theta} \right)^r \left(\frac{\theta}{\lambda_{it} + \theta} \right)^\theta \qquad (4.7)$$

In this equation λ_{it} is the expected value of y_{it}, θ is the overdispersion parameter, and $\Gamma(\cdot)$ is the gamma function. As $\theta \to \infty$, this distribution converges to the Poisson distribution. As with the Poisson model, we assume that the expected value of y_{it} is described by a log-linear regression:

$$\log \lambda_{it} = \mu_t + \beta\mathbf{x}_{it} + \gamma\mathbf{z}_i + \alpha_i \tag{4.8}$$

where the α_i are treated as fixed effects. Conditional on α_i, the several event counts for each individual (a firm in our example) are assumed to be independent, although unconditionally they may be dependent.

How can this model be estimated? Unlike the Poisson model, conditional likelihood is not an option. In technical terminology, the total count for each individual is not a "complete sufficient statistic" for α_i, so conditioning on the total does not remove α_i from the likelihood function. Hausman, Hall, and Griliches (1984) proposed a rather different fixed effects negative binomial regression model, and they derived a conditional ML estimator for that model. In fact, their method has been incorporated into the **xtnbreg** command in Stata. But Allison and Waterman (2002) have shown that this is not a true fixed effects regression model, and the method does not, in fact, control for all stable predictors, as we will see below.

Instead, we will do unconditional ML by estimating negative binomial regression models that include dummy variables for all individuals (except one). In Stata, this can be done with the **nbreg** command.[5] Computation for the model is quite slow because of the many coefficients that must be estimated for the firm dummy variables. To speed things up a little, I omitted the 22 firms that had no patents in any of the 5 years. These firms contribute nothing to the likelihood function and their dummy variable coefficients do not converge.

The results in Table 4.6 should be compared with those for the fixed effects Poisson regression in Table 4.3. Coefficients for the dummy variables for firms are not shown. It's apparent that the coefficients for the negative binomial model are very similar to those for the Poisson model. Moreover, the standard errors and test statistics for the negative binomial model are close to those for the Poisson model with bootstrapped standard errors. The estimated parameter labeled Alpha is a measure of overdispersion. It's actually an estimate of $1/\theta$, where θ is the parameter in Equation 4.7. At more than 10 times its standard error, Alpha is clearly greater than 0, which means that there is a significant amount of overdispersion.

Table 4.6   Unconditional Estimates of Fixed Effects Negative Binomial Model

	Fixed Effects		
	Coefficient	Conventional Standard Error	Outer Product of Gradient (OPG) Standard Error
RD0	0.371**	0.063	0.072
RD1	−0.083	0.068	0.073
RD2	0.064	0.064	0.075
RD3	0.014	0.060	0.071
RD4	0.034	0.056	0.060
RD5	0.002	0.046	0.052
TIME_2	−0.049*	0.023	0.027
TIME_3	−0.051*	0.023	0.029
TIME_4	−0.159**	0.024	0.028
TIME_5	−0.224**	0.025	0.028
Constant	3.677	0.118	0.101
Alpha	0.020**	0.002	0.002

*$.01 < p < .05$.

**$p < .01$.

Stata also reports a likelihood ratio chi-square statistic of the null hypothesis that Alpha = 0, which, in this case, has a value of 499.54 with 1 degree of freedom, statistically significant by any standard. This statistic is calculated by taking twice the difference between the log-likelihood for the negative binomial model and the log-likelihood for the Poisson. The reason this works is that the Poisson is a special case of the negative binomial such that Alpha is equal to 0. The implication is that we should reject the Poisson model in favor of the negative binomial model.

The negative binomial model clearly fits these data better than the Poisson model. But unlike the Poisson model (where conditional and unconditional estimates must be identical), there is no guarantee that unconditional negative binomial estimation is resistant to the biases that arise from the incidental parameters problem (discussed for the logistic model in Chapter 3). Using Monte Carlo simulations, Allison and Waterman (2002) found that that the unconditional negative binomial estimator did not show any substantial bias from incidental parameters. They also showed that negative binomial estimators had substantially smaller true standard errors than Poisson estimators. Unconditional negative binomial estimation did have one flaw, however: Confidence intervals tended to be too small (although the undercoverage was not nearly as severe as for the Poisson model). Under many conditions, the nominal 95% confidence intervals covered the true value only about 85% of the time. This problem was easily corrected by adjusting the standard errors for overdispersion using a formula based on the deviance statistic. When this was done in simulations, the actual coverage rates were very close to the nominal 95% confidence intervals for nearly all conditions. Although Stata does not report the deviance statistic required for this correction, I have found that the standard errors produced by the **vce(opg)** option are about the same as those produced by the deviance correction. These standard errors are shown in the third column of Table 4.6.

Computation time for the unconditional negative binomial estimates was tolerable for the patent data, but it could easily become a problem for very large data sets with lots of dummy variable coefficients to estimate. Greene (2001) has shown how such computational difficulties can be readily overcome, but that would require modification of the existing Stata algorithms.

Earlier, I stated that the conditional negative binomial method embodied in Stata's **xtnbreg** command was not a true fixed effects method. Table 4.7 provides an illustration of that fact. These estimates were produced using **xtnbreg** with the fixed effects option. For Model 1, which includes only time-varying predictors, the results are quite similar to what we saw in Table 4.6 for the unconditional method. But note that we get an estimate for the constant, which would be expected to drop out of a conditional likelihood function. Model 2 includes two time-invariant predictors, SCIENCE and LOGSIZE. If the conditional likelihood were really controlling for all time-invariant predictors, then we shouldn't be able to include these variables because they would be redundant. In addition, we see that LOGSIZE has a highly significant coefficient, and that the effect of RD0 changes with the introduction of SCIENCE and LOGSIZE. None of this makes sense for a true fixed effects estimator.

Table 4.7 Estimates for Stata's "Fixed Effects" Negative Binomial Model

	Model 1		Model 2	
	Coefficient	Standard Error	Coefficient	Standard Error
RD0	0.319**	0.067	0.273**	0.071
RD1	−0.080	0.077	−0.098	0.077
RD2	0.056	0.071	0.032	0.071
RD3	−0.013	0.066	−0.020	0.066
RD4	0.035	0.062	0.016	0.063
RD5	0.009	0.052	−0.010	0.053
TIME_2	−0.042	0.025	−0.038**	0.024
TIME_3	−0.049	0.025	−0.040**	0.025
TIME_4	−0.161**	0.026	−0.144**	0.026
TIME_5	−0.215**	0.026	−0.196**	0.027
SCIENCE			0.018	0.198
LOGSIZE			0.207**	0.078
Constant	2.424**	0.175	1.661**	0.343

$**p < .01.$

A Hybrid Approach

As we saw in the previous chapters, it's possible to combine the fixed effects and random effects approaches to get some of the virtues of each. Within this framework, we can perform a statistical test of the fixed effects versus random effects model, and we can estimate effects of variables that do not change over time. As before, the first step is to calculate the mean of each time-varying predictor variable for each individual, then calculate the deviations from those means. The next step is to run a regression model with both the deviation variables and the mean variables as predictors. Here, we will estimate negative binomial regression models because they are less prone to overdispersion. To get correct standard errors, it's important to use an estimation method that allows for dependence among the multiple observations for each individual. Either a random effects model or a population-averaged (GEE) model can accomplish that.[6]

I estimated both the random effects model and the GEE (population-averaged) model with Stata's **xtnbreg** command, with the results shown in Table 4.8. For GEE, I estimated the default "exchangeable" model that assumes equal correlations among all years within firms, making it essentially equivalent to the random effects model. All the variable names beginning with D refer to the deviation variables while all those beginning with M are firm-specific means.

Table 4.8 Hybrid Estimates for Negative Binomial Model

	Random Effects		GEE	
	Coefficient	Standard Error	Coefficient	Standard Error
DRD0	0.322**	0.071	0.410**	0.120
DRD1	−0.057	0.076	−0.129	0.120
DRD2	0.081	0.068	0.056	0.082
DRD3	−0.006	0.064	−0.012	0.095
DRD4	0.011	0.059	0.007	0.099
DRD5	0.019	0.050	−0.062	0.088
MRD0	−0.336	0.697	0.031	0.798
MRD1	2.246	1.426	1.080	1.722
MRD2	−1.985	1.585	−1.110	1.850
MRD3	−0.500	1.408	−0.075	1.566
MRD4	1.248	1.106	1.119	1.136
MRD5	−0.051	0.517	−0.274	0.478

	Random Effects		GEE	
	Coefficient	Standard Error	Coefficient	Standard Error
SCIENCE	0.057	0.103	−0.007	0.112
LOGSIZE	0.119**	0.045	0.105*	0.052
TIME_2	−0.042*	0.021	−0.052	0.034
TIME_3	−0.049*	0.022	−.049	0.040
TIME_4	−0.168**	0.023	−0.100*	0.047
TIME_5	−0.208**	0.025	−0.209**	0.050
Constant	1.038**	0.171	1.002	0.178

*.01 < p < .05.

**p < .01.

The coefficients for the deviation variables can be interpreted as fixed effects estimates in the sense that they are based only on within-firm variation and, hence, control for all stable predictors. In fact, they are quite close to the fixed effects coefficients of the R&D variables in Table 4.6. As in that table, the only deviation variable that achieves statistical significance is DRD0, the log of research and development expenditures in the current year. The GEE coefficient of 0.41 says that a 1% increase in R&D is associated with a 0.41% increase in the number of patents.

As usual, one of the attractions of the hybrid method is the ability to include time-invariant predictors, in this case SCIENCE and LOGSIZE. The latter has a significant positive effect on number of patents. Keep in mind, however, that these coefficients do not control for omitted explanatory variables, unlike the deviation coefficients.

The other attraction of the hybrid model is the ability to test the fixed effects model against the more restricted random effects model. This is accomplished by testing whether the deviation coefficients are the same as the corresponding mean coefficients. We see in Table 4.8 that the coefficients for the means are generally quite different from the deviation coefficients, although none of the mean coefficients is statistically significant. Chi-square tests for the difference provide only marginal evidence in favor of fixed effects. For the random effects model, the Wald chi-square was 12.16 with 6 df ($p = .06$). For the GEE model, the Wald chi-square was 12.87 with 6 df ($p = .04$).

Summary

Fixed effects regression models for count data can be estimated under the assumption that the dependent variable has either a Poisson or negative binomial distribution. When there are only two periods per individual, conditional ML estimation of a fixed effects Poisson model can be implemented by transforming the Poisson model into a logistic regression model for grouped data, with difference scores for predictor variables. When there are more than two observations per individual, conditional ML estimation of the Poisson model can be accomplished with Stata's **xtpoisson** command.

Unconditional ML can be done with standard Poisson regression software, using dummy variables to represent the fixed effects. Unlike logistic regression, conditional and unconditional estimation of the fixed effects Poisson model produce identical coefficients and standard errors. Unfortunately, the standard errors are often severely biased due to overdispersion. In Stata, I used bootstrap standard errors to correct for overdispersion, but some software packages have computationally simpler methods.

A better approach to overdispersion is to estimate negative binomial regression models with an overdispersion parameter. Such models cannot be estimated by conditional ML, however. Unconditional ML can be done with any negative binomial regression software using dummy variables for the fixed effects.

The hybrid method allows for the estimation of fixed effects coefficients for time-varying predictors while also estimating the effects of time-invariant predictors. As we saw in Chapters 2 and 3, each time-varying predictor is decomposed into two parts: an individual-specific mean and a deviation from that mean. The regression model includes both sets of variables, along with any time-invariant predictors. Within-individual dependence can be handled by either GEE estimation or ML estimation of a random effects model.

Notes

1. This chapter does not consider zero-inflated Poisson and negative binomial models for three reasons: They are much more complex, there is little software available for panel data, and, finally, the negative binomial model itself often provides a satisfactory fit to data with large numbers of zero counts.

2. The **reshape** command in Stata makes it easy to restructure the data in this way.

3. Because the bootstrap method involves random draws, the bootstrap standard errors will differ slightly from one run to another. The degree of variability can be reduced by increasing the number of bootstrap samples.

4. The GEE estimates in Table 4.3 were obtained using an "unstructured" correlation matrix for the counts in the 5 years.

5. The **nbreg** command can fit two different versions of the negative binomial model. In the default version (the one preferred here), the variance is a function of the mean. Alternatively, using the **dispersion(constant)** option, the variance can be specified as constant. Although that may seem attractive, it's not really sensible for the kinds of applications envisioned here.

6. Unlike the logistic model in Chapter 3, there does not appear to be any distinction between population-averaged models and subject-specific models for the negative binomial model. That implies that the random effects estimates should not be any larger in magnitude than the GEE estimates.

CHAPTER 5. FIXED EFFECTS MODELS FOR EVENTS HISTORY DATA

Event history analysis is the name given to a set of statistical methods that are designed to describe, explain, or predict the occurrence of events. Outside the social sciences, these methods are commonly called *survival analysis*, owing to the fact that they were first developed by biostatisticians to analyze the occurrence of deaths. But it so happens that these same methods are perfectly appropriate for a vast array of social phenomena such as births, marriages, divorces, job terminations, promotions, arrests, migrations, and revolutions. There also many other names for event history methods, including failure time analysis, hazard analysis, transition analysis, and duration analysis.

In general, an event may be defined as a *qualitative* change that occurs at some particular point in time. To apply event history methods you need event history data, which is simply a longitudinal record of when events occurred to some individual or sample of individuals. For example, if you ask a sample of women to report the birth dates of all their children, you will get a set of event history data that will allow you to analyze the occurrence of births. Of course, if you want to do a causal or predictive analysis, you will also want to measure possible explanatory variables, such as the woman's date of birth, education, family income, marital status, and so on.

Let's make this example more concrete. In the 1995 National Survey of Family Growth (NSFG), a representative sample of American women were asked to report information on the births of all children ever born to them (www.cdc.gov/nchs/nsfg.htm). The subsample of the data used here included 6,911 women who experienced at least one birth. These women reported a total of 14,932 live births. For each of these births, I calculated a birth interval, labeled DUR: the length of time (in months) from the current birth to the next birth, or until the interview date if no subsequent birth was observed. Potential predictors of these birth intervals include several variables that characterize the current birth:

PREGORDR	Order of the birth (1 through 15)
MARRIED	1 if married at the time of the birth, otherwise 0
AGE	Mother's age (in years) at birth
PASST	1 if delivery was paid for, in part, by public assistance funds, otherwise 0

NOBREAST   1 if mother did not breast-feed baby, otherwise 0

LBW   1 if low birth weight, otherwise 0

CAESAR   1 if birth was by Caesarian section, otherwise 0

MULTIPLE   1 if more than one baby born, otherwise 0

There is also a variable COLLEGE equal to 1 if the woman had some college education (at the time of the interview), otherwise 0, and a variable BIRTH equal to 1 if the interval ended in another birth or 0 if the interval was terminated by the interview—a censored interval. The data set had 6,911 censored birth intervals in this data set. Every woman had exactly one censored interval because her last interval was terminated by the interview. Last, the variable CASEID is an ID number that is common to all the birth intervals for the same woman. Our goal is to estimate a regression model for the length of birth interval.

Cox Regression

The most popular method for analyzing event history data is Cox regression, named after David Cox (1972) who developed the *proportional hazards model* and the *partial likelihood* method for estimating that model. Before proceeding to a fixed effects analysis, I briefly review this method.

Rather than directly modeling the length of the interval, the dependent variable in Cox regression is the *hazard* or instantaneous likelihood of event occurrence. For repeated events, the hazard may be defined as follows. Let $N_i(t)$ be the number of events that have occurred to individual i by time t. The hazard for individual i at time t is given by

$$h_i(t) = \lim_{\Delta t \to 0} \frac{\Pr[N_i(t + \Delta t) - N_i(t) = 1]}{\Delta t} \qquad (5.1)$$

In words, this equation says that we should consider the probability of one additional event in some small interval of time Δt. Then form the ratio of this probability to Δt, and take the limit of this ratio as Δt goes to 0. For repeated events, the hazard function is also known as the intensity function.

Next, we model the hazard as a function of the predictor variables. Letting $h_{ik}(t)$ be the hazard for the kth event for individual i, a proportional hazards model is given by

$$\log h_{ik}(t) = \mu(t - t_{i(k-1)}) + \beta \mathbf{x}_{ik} \qquad (5.2)$$

where x_{ik} is a column vector of predictor variables that may vary across individuals and across events, β is a row vector of coefficients, $t_{i(k-1)}$ is the time of the $(k-1)$th event, and $\mu(\cdot)$ is an unspecified function of the length of time since the most recent event. In this model, we assume that $\mu(\cdot)$ is the same function for all individuals in the sample.

A remarkable thing about partial likelihood is that it can estimate β without making any assumptions about the function μ. For details on how this is accomplished, see Allison (1995). In Stata, Cox regression is implemented with the **stcox** command. Table 5.1 (first two columns of numbers) gives the results of estimating a model for the birth interval data, treating all the birth intervals as independent observations, that is, as if each interval came from a different woman in the population. All the variables except for low birth weight have highly significant effects on the hazard for the next birth. Women who are married or on public assistance have higher hazards for a birth. All the other variables have negative coefficients.

To get a more precise interpretation of these results, it's helpful to look at the last column (labeled "hazard ratio") containing the exponentiated values of the parameter estimates. Hazard ratios are interpreted almost

Table 5.1 Cox Regression Estimates for a Conventional Model[a]

	Coefficient	Conventional Standard Error	Robust Standard Error	Hazard Ratio
PREGORDR	−0.163	0.011	0.016	0.849
AGE	−0.065	0.003	0.003	0.937
MARRIED	0.221	0.029	0.030	1.247
PASST	0.137	0.029	0.029	1.147
NOBREAST	−0.270	0.023	0.023	0.763
LBW	−0.003	0.042	0.043	0.997
CAESAR	−0.116	0.030	0.028	0.890
MULTIPLE	−0.702	0.143	0.144	0.495
COLLEGE	−0.207	0.026	0.026	0.813

a. All coefficients have p values less than .01, except for LBW, which has a p value above .90.

exactly like odds ratios in logistic regression. For example, MARRIED has a hazard ratio of 1.25. This means that women who are married at the time of a birth have a hazard for another birth that is 25% larger than the hazard for unmarried women (controlling for other variables in the model). The hazard ratio for MULTIPLE is 0.493, which means that if a woman has twins, the hazard for another birth is cut in half. For AGE, the hazard ratio is 0.936, which means that each additional year of the mother's age *reduces* the hazard of a subsequent birth by $100(1 - 0.936) = 6.4\%$.

Unfortunately, these results are potentially problematic. Sixty-nine percent of the women contributed at least two birth intervals to the data set, and it's reasonable to suspect that there would be some dependence among these repeated observations. In particular, it's natural to suppose that some women have persistently short birth intervals while others have persistently long intervals. The failure to address this dependence could lead to serious underestimates of the standard errors and p values.

Fortunately, it's easy to correct the standard errors using the method of robust variance estimation that we have used in the previous chapters (Therneau & Grambsch, 2000). Robust standard errors, obtained with the option `vce(cluster caseid)`, are shown in the third column of numbers in Table 5.1. For the most part, the corrections here are rather small. One exception is the corrected standard error for PREGORDER, which is 37% larger than the uncorrected version. This produces a corrected z statistic that is only about half the uncorrected statistic, although still highly significant.

Cox Regression With Fixed Effects

We are now ready to incorporate fixed effects into the Cox regression model. As usual, this allows us to control for all stable predictor variables, while fixing the problem of dependence among the repeated observations. As in previous fixed effects models, α_i represents the combined effects of all stable predictor variables. The first version of our fixed effects regression model is

$$\log h_{ik}(t) = \mu(t - t_{i(k-1)}) + \beta x_{ik} + \alpha_i \tag{5.3}$$

How can Equation 5.3 be estimated for our birth interval data? An obvious possibility is to put dummy variables in the model for all women (except one). This method worked well for linear models and for Poisson and negative binomial regression models, but it runs into serious difficulties here. For one thing, there is the practical problem of estimating a Cox regression with 6,910 dummy variables.[1]

A more fundamental difficulty is the possible bias introduced by estimating so many "incidental parameters." In the previous chapters, we found that this bias could be quite serious for logistic regression models but not for Poisson or negative binomial models. Elsewhere (Allison, 2002), I have shown that Cox regression is more like logistic regression in this respect. When the average number of intervals per person is less than three, using dummy variables to implement fixed effects produces regression coefficients that are biased (away from zero) by approximately 30% to 90%, depending on the level of censoring (a higher proportion of censored cases produces greater inflation).

Fortunately, there is an alternative method that is easily implemented and very effective. It is similar to the conditional likelihood methods used for both logistic and Poisson regression because the coefficients for the dummy variables are not actually estimated but are eliminated from the likelihood function. First, we modify Equation 5.3 by defining

$$\mu_i(t-t_{i(k-1)}) = \mu(t-t_{i(k-1)}) + \alpha_i$$

which yields

$$\log h_{ik}(t) = \mu_i(t-t_{i(k-1)}) + \beta \mathbf{x}_{ik} \tag{5.4}$$

In this equation, the fixed effect α_i has been absorbed into the unspecified function of time, which is now allowed to vary from one individual to another. Note that the only difference between Equation 5.4 and the conventional Cox model in Equation 5.2 is the i subscript on μ. Thus, each individual has her own hazard function, which is considerably less restrictive than allowing each individual to have her own constant.

Equation 5.4 can be estimated by standard Cox regression programs with the widely available option of stratification. Stratification allows different subgroups to have different baseline hazard functions, while constraining the coefficients to be the same across subgroups. It is accomplished by constructing a partial likelihood function for each subgroup, multiplying those likelihood functions together, and then maximizing the resulting likelihood function with respect to the coefficient vector β. With the **stcox** command in Stata, stratification is implemented by specifying the option **strata(caseid)**, which means that each of the 6,911 women is treated as a separate stratum. That may seem like an enormous number of strata, but **stcox** handles them with ease.

The results in Table 5.2, Model 1, show some noteworthy differences from those in Table 5.1. First, there's nothing reported for COLLEGE. Like most of our fixed effects methods (except for the hybrid method), we can't estimate

Table 5.2 Cox Regression Estimates for Fixed Effects Models

	Model 1			Model 2	
	Coefficient	Standard Error	Hazard Ratio	Coefficient	Standard Error
PREGORDR	−0.711**	0.034	0.491	−0.712**	0.034
AGE	0.007	0.011	1.007	0.007	0.011
MARRIED	0.181**	0.070	1.199	0.182**	0.070
PASST	0.077	0.069	1.080	0.076	0.069
NOBREAST	−0.128*	0.060	0.879	0.043	0.100
LBW	−0.237**	0.081	0.789	−0.243**	0.081
CAESAR	−0.079	0.093	0.923	−0.080	0.093
MULTIPLE	−0.607**	0.218	0.545	−0.590**	0.219
COLLEGE	(drop)			(drop)	
COLLBREAST				−0.267*	0.125

*.01 < p < .05.

**p < .01.

coefficients for variables that do not vary within person. Moving upward from COLLEGE, we see that the coefficient for multiple birth is about the same as the previous estimates. But the coefficient for CAESAR is somewhat attenuated and no longer statistically significant. Low birth weight was previously far from statistically significant, but here the p value is less than .01. The hazard ratio for LBW tells us that a low birth weight is associated with a 21% reduction in the hazard for a subsequent birth. The effect of breast-feeding is attenuated, both in magnitude and in significance. Public assistance was previously highly significant, but here it's not significant at all. The effect of marital status is about the same. Age is no longer statistically significant. On the other hand, the effect of pregnancy order is *much* greater, both in magnitude and in statistical significance. Each additional birth is associated with about a 50% reduction in the hazard for a subsequent birth.

Why are the fixed effects estimates so different from the conventional Cox regression estimates? Like any fixed effects method, this one controls for all stable predictors, so it's possible that some of the earlier results in

Table 5.1 were spurious. If I had to choose between the conventional results in Table 5.1 and the fixed effects results in Table 5.2, I would emphatically choose the latter. The thing to keep in mind is that, in this analysis, each woman is being compared with herself in a different birth interval. For each woman, we're asking why some of her birth intervals are longer or shorter than others. Is it, for example, because she's married for some of the intervals and not for others? This approach will produce different answers than asking why some women tend to have longer birth intervals than others.

This aspect of the fixed effects method is especially relevant to the PREGORDR variable. In a conventional Cox regression, this variable is likely to have a spuriously positive effect on the hazard. For a fixed interval of time, women who have more births in that interval will necessarily have shorter birth intervals. By doing a fixed effects analysis, we are able to remove that artifact, which is why the negative coefficient becomes so much larger than before.

As with linear and logistic models, even though the fixed effects Cox model will not estimate the effects of time-invariant predictors such as COLLEGE, it is possible to estimate interactions between time-invariant variables and other variables. For example, we can estimate a model that includes an interaction between COLLEGE and NOBREAST. This is accomplished by simply including a predictor variable that is the product of COLLEGE and NOBREAST. Results are in Table 5.2, Model 2. The interaction is statistically significant at the .03 level. But how is it interpreted? The "main effect" of NOBREAST represents the effect of this variable when COLLEGE = 0, that is, among women without a college education. That coefficient is positive and far from statistically significant. The effect of NOBREAST among college-educated women is found by adding the main effect to the interaction $(-0.2659 + 0.0421) = -.22$. Using the **test** command, one can show that this sum is significantly different from 0. The conclusion is that breast-feeding increases the hazard of a subsequent birth among college-educated women, but not among other women.

Stata can also estimate a random effects Cox model, as specified by Equation 5.3 with the assumption that α_i has a gamma distribution and is independent of $\mathbf{x}_i$. Models of this sort are often called "shared frailty" models, with α_i (or its exponential transform) described as the frailty term. The idea is that some individuals are more frail than others and, hence, are more likely to experience the event. The **stcox** option for estimating such models is **shared(caseid)**. Unfortunately, Stata suffered a computational failure in attempting to estimate this model for the birth-interval example, apparently because of the size of the sample.

Computations *were* successful for a random effects Gompertz model (implemented with the **streg** command), which is a special case of Equation 5.3, whereby $\mu(\cdot)$ is specified as a linear function. Estimates for that model are shown in Table 5.3. Results are remarkably similar to the conventional Cox model estimates in Table 5.1, even without robust standard error corrections. This should not be surprising given that the estimated variance of α_i was not significantly different from zero. Of course, the fixed effects estimates in Table 5.2 are quite different from the random effects estimates, demonstrating again that controlling for unobserved heterogeneity can be important even when a random effects model provides no evidence for the existence of such heterogeneity.

Some Caveats

Despite the attractions of fixed effects Cox regression, it also has the usual disadvantages. As with other fixed effects methods, there may be a substantial loss of power compared with the conventional analysis. In this example, any woman with only one birth interval gets excluded because

Table 5.3   Estimates for a Random Effects Gompertz Model[a]

	Coefficient	Standard Error	Hazard Ratio
PREGORDR	−0.163	0.011	0.850
AGE	−0.061	0.003	0.941
MARRIED	0.200	0.029	1.222
PASST	0.132	0.029	1.141
NOBREAST	−0.265	0.023	0.767
LBW	0.007	0.042	1.007
CAESAR	−0.095	0.030	0.910
MULTIPLE	−0.675	0.143	0.509
COLLEGE	−0.186	0.026	0.830

a. All coefficients have *p* values less than .01, except for LBW, which has a *p* value above .90.

that interval can't be compared with any others. This eliminates 2,109 birth intervals. Second, among women with exactly two birth intervals, if the second interval (which is always right censored) is shorter than the first, both intervals will be excluded. Here's why. Suppose the first interval is 28 months long and the second interval is censored at 20 months. In constructing the partial likelihood for the birth that occurs at 28 months, the algorithm looks for other intervals (from the same woman) that are "at risk" of the event at that same time. But the other interval was censored at 20 months, so for that interval, the woman is no longer at risk of an observable birth at 28 months. As a result, there is nothing with which to compare the birth, and the woman is eliminated from the partial likelihood function. For the NSFG data, the elimination of these intervals results in the loss of another 1,468 cases.

Finally, even for those observations that are retained, the fixed effects method essentially discards information about variation *across* women and only uses variation *within* women. So if a particular covariate varies a great deal across women, but shows little variation over time for each woman, the coefficient for that variable will be unreliably estimated. The variable PASST, for example, has 80% of its variance across women and only 20% within women. Not surprisingly, the standard error for its coefficient is more than twice as large in Table 5.2 as compared with Table 5.1, which was based on variations both within and between women.

Besides the usual limitations of fixed effects methods, fixed effects Cox regression is also susceptible to bias for certain kinds of variables. These problems are most likely to occur with the kind of data structure that occurs in the birth interval study. In that structure, individuals are observed for a fixed period of time and may have multiple events during that period, but only the last interval is censored. Chamberlain (1985) argued that this structure violates a basic condition of likelihood-based estimation because the probability that an interval is censored depends on the length of the previous intervals.

In a simulation study (Allison, 1996), I showed that this violation does not create a serious problem for most predictor variables, but could lead to biases in estimating the effects of variables that describe the previous event history. In particular, fixed effects partial likelihood tends to find negative effects on the hazard for the number of previous events and the length of the previous interval, even when those variables do not have true effects. This is certainly consistent with the results in Table 5.2, which show strong negative effects of pregnancy order on the hazard of a subsequent birth. This problem tends to be most severe when the average number of events per individual is low, and the proportion of intervals that are censored is high. On the other hand, I argued earlier that a conventional Cox regression is likely to be even more biased for the effect of number of prior events, but in the opposite direction.

The Hybrid Method for Cox Regression

In the previous chapters, we saw that we could duplicate or closely approximate the results from a fixed effects analysis by decomposing the time-varying predictors into individual-specific means and deviations from those means, and then putting all the variables into a conventional regression analysis, possibly correcting for dependence among the multiple observations for each individual. Unfortunately, for reasons that are not clear, this approach does not seem to work well for Cox regression. For example, if the hybrid method is applied to the birth interval data, several of the variables have coefficients and p values that are dramatically different from those shown in Table 5.2. My simulation studies of the hybrid method for Cox regression have also been discouraging. Accordingly, I cannot recommend the hybrid method for event history analysis.

Fixed Effects Event History Methods for Nonrepeated Events

Fixed effects Cox regression requires that at least some of the individuals in a sample experience more than one event so that within-individual comparisons are possible. Obviously, then, the method cannot be applied to a nonrepeatable event such as death. Nevertheless, under certain conditions, it may be possible to do a fixed effects analysis for nonrepeatable events by treating time as discrete and applying conditional logistic regression. In the epidemiological literature, this type of analysis is called a *case-crossover study* (Maclure, 1991), although the implementation I describe here is a little different from the way in which epidemiologists usually do it.

As usual, I begin with an empirical example. Suppose we want to answer the following question: Does the death of a wife increase the hazard for the death of her husband? That's a difficult question to answer with confidence because any association between husband's death and wife's death could be due to the effects of common environmental characteristics. Most of them will have lived in the same house in the same neighborhood for substantial periods of time. Moreover, they will tend to have come from similar social and economic backgrounds and have similar lifestyles. Unless we can control for those commonalities, any observed association between the death of one spouse and the other is likely to be spurious. Hence, a fixed effects analysis is highly desirable as a way to control for stable, unmeasured predictors.

To answer this question, we shall analyze data on 49,990 married couples in which both spouses were alive and at least 68 years old on January 1, 1993.[2] Death dates for both spouses are available through May 30, 1994. During that 17-month interval, there were 5,769 deaths of the husband and 1,918 deaths of the wife. We regard time as consisting of discrete units, in this case days, which we can enumerate $t = 1, 2, 3, \ldots$ Let p_{it} be the probability that husband i dies on day t, given that he was still alive on the preceding day, and let $W_{it} = 1$ if the wife i was alive on day t, otherwise 0.

We will represent the effect of wife's vital status on the probability of husband's death by a logistic regression model

$$\log\left(\frac{p_{it}}{1 - p_{it}}\right) = \alpha_i + \gamma t + \beta W_{it} \tag{5.5}$$

where γt represents a linear effect of time on the log-odds of death and α_i represents the fixed effects of all unmeasured variables that are constant over time. Note that no time-invariant predictors are included in the model because their effects are absorbed into the α_i term.

We now try to estimate this model by the method of conditional maximum likelihood, described in Chapter 3, which eliminates the α_is from the estimating equations. Here's how the data set is constructed. For men who died, a separate observational record is created for each day that the couple is observed, from Day 1 (January 1, 1993) until the day of death. For each of these couple-days, the dependent variable Y_{it} is coded 0 if the man remained alive on that day, and coded 1 if he died on that day. Thus, a man who died on June 1, 1993, would contribute 152 couple-days; 151 of those would have a value of 0 on Y_{it}, while the last would have a value of 1. The predictor variable W_{it} is coded 0 for all days on which the wife was alive and 1 for all days on which she was dead. No observations are created for men who did not die because, in a fixed effects analysis for dichotomous outcomes, individuals who do not change contribute nothing to the likelihood function. The total number of couple-days for the working data set is 1,377,282. The model can then be estimated with the Stata commands **xtlogit** or **clogit**, as described in Chapter 3.

Unfortunately, for either of these commands, the algorithm used to maximize the likelihood function does not converge. The log-likelihood quickly goes to 0 and the iteration sequence continues without end. The reason for the convergence failure is that each couple's sequence of observations consists of a string of 0s on the dependent variable, followed by a 1. That is, the event always occurs at the last observation unit. As a consequence, time or any monotonically increasing function of time (such as the logarithm of time or the square root of time) will perfectly predict the outcome for that couple, making it impossible to get maximum likelihood estimates for that covariate or any other covariate in the model. In the logistic regression literature, this problem is known as *complete separation* (Albert & Anderson, 1984; Allison, 2004).[3]

Actually, for our mortality example, the problem of nonconvergence is not confined to the time variable. If we remove time from the model, we still get nonconvergence (although the problem now is not complete separation but quasi-complete separation). Because W_{it}, the dummy variable for wife's death, may increase with time but never decrease, it perfectly predicts the occurrence of a death on the last day. Consequently, its coefficient gets larger at each iteration of the algorithm.

One way to circumvent this problem is to redefine W_{it} to be an indicator of whether the wife died within, say, the previous 60 days. This covariate changes from 0 to 1 when the wife dies, but then goes back to 0 after 60 days (if the husband is still alive). Estimating the model with varying windows of time can give useful information about how the effect of wife's death starts, peaks, and stops.

The upper panel of Table 5.4 gives fixed effects estimates of the odds ratios for the effect of wife's death on husband's death using several different windows of time (but no effect of time itself). In all cases, the odds ratios exceed 1.0, and are statistically significant for the 60-day interval and the 30-day interval. For the latter, the odds of husband's death on a day in which the wife died during the previous 30 days are nearly double the odds if the wife did not die during that interval. The lower panel of Table 5.4 gives odds ratios from conventional logistic regression, that is,

Table 5.4 Odds Ratios for Predicting Husband's Death From Wife's Death Within Varying Intervals of Time

	Wife Died Within				
	15 Days	30 Days	60 Days	90 Days	120 Days
Fixed Effects Estimates					
Odds ratio	1.26	1.96	1.61	1.27	1.26
p Value	.54	.006	.03	.24	.25
Conventional Estimates					
Odds ratio	1.13	1.56	1.21	0.97	0.93
p Value	.71	.04	.29	.87	.61

with no controls for stable, unobserved covariates. The odds ratios are all smaller for these estimates compared with those in the upper panel, and the p values are all higher.

Although these results are intriguing, the danger is that there is no control for change over time. This is not merely a technical problem, but one that can seriously compromise any conclusions drawn from a case-crossover study (Greenland, 1996; Suissa, 1995). For our example, if there is *any* tendency for the incidence of wife's death to increase over the period of observation, this can produce a spurious relationship between wife's death (however coded) and husband's death. Intuitively, the reason is that husband's death always occurs at the end of the sequence of observations for each couple, so any variable that tends to increase over time will appear to increase the hazard of husband's death.

We now consider an alternative fixed effects method that appears to solve the problems that arise from uncontrolled dependence on time. Introduced by Suissa (1995) who called it the "case-time-control" design, the key innovation in this approach is the computational device of reversing the dependent and independent variables in the estimation of the conditional logit model. This makes it possible to introduce a control for time, something that cannot be done with the case-crossover method.

As is well-known, when both the dependent and the independent variables are dichotomous, the odds-ratio is symmetric—reversing the dependent and independent variables yields the same result, even when there are other predictors in the model.[4] In the case-time-control method, the working dependent variable is the dichotomous covariate—in our case, whether or not the wife died during the preceding specified number of days. Independent variables are the dummy variable for the occurrence of an event (husband's death) on a given day and some appropriate representation of time, for example, a linear function. Again, a conditional logistic regression is estimated with each couple treated as a separate stratum. Under this formulation, there is no problem including time as a covariate because the working dependent variable is not a monotonic function of time.

In Suissa's formulation of the method, it is essential to include data from all individuals, both those who experienced the event and those who are censored. However, his model was developed for data with only two points in time for each individual, an event period and a control period. In that scenario, the covariate effect and the time effect are perfectly confounded if the sample is restricted to those who experienced events. On the other hand, censored individuals provide information about the dependence of the covariate on time, information that is not confounded with the occurrence of the event.

In contrast, our data set (and presumably many others) has multiple "controls" at different points in time for each individual. That eliminates the complete confounding of time with the occurrence of the event (husband's death), making it possible to apply the case-time-control method to uncensored cases only. That's a real boon in situations where it is difficult or impossible to get information for those who did not experience the event. The only restriction is that when the model is estimated without the censored cases, one cannot estimate a model with a completely arbitrary dependence on time, that is, with dummy variables for every point in time.

Of course, if the censored cases are available (as in our data set), more precise estimates of the dependence on time can be obtained by including them. But even if censored cases are available, there is a potential advantage to limiting the analysis to those who experienced the event. The case-time-control method has been criticized for assuming that the dependence of the covariate on time is the same among those who did and did not experience the event (Greenland, 1996). This criticism has no force if the data are limited to those individuals who experience the events.

For the mortality data, the working data set is the same as before with one record for each day of observation, from the origin until the time of husband's death or censoring. Because conditional logistic regression requires variation on the dependent variable for each conditioning stratum, we can eliminate couples whose wife did not die before the husband, with no loss of information.

The working data set has 39,942 couple-days, which came from only 126 couples. This is the number of couples in which the husband died *and* the wife died before the husband. Although this is a tiny fraction of the original sample of 49,990 couples, it's the only group that contains information about the effect of wife's death on husband's death using a fixed effects approach. Is that a problem? Well, not if the same model (with the same coefficients) applies to everyone in the population. But if the model differs across subgroups, the results obtained for these 126 couples will accurately describe *them*, but not the population as whole.

The working model is defined as follows. Let H_{it} be a dummy variable for the death of the husband i on day t, and let P_{it} be the probability that the wife's death occurred within a specified number of days prior to day t. The logistic regression model is

$$\log\left(\frac{P_{it}}{1 - P_{it}}\right) = \alpha_i + \beta_1 H_{it} + \beta_2 t + \beta_3 t^2 \tag{5.6}$$

Table 5.5   Odds Ratios for Predicting Husband's Death From Wife's Death Within Varying Intervals of Time, Case-Time-Control Method

	15 Days	30 Days	60 Days	90 Days	120 Days
Odds ratio	1.26	2.08	1.74	1.28	1.11
p Value	.54	<.004	.01	.25	.63

This model allows for a quadratic dependence on time, although other functions could be used instead.

Table 5.5 gives estimates of the odds ratios for varying windows of time. Results are quite similar to those in Table 5.4, which used the case-crossover method. Again, the evidence suggests that the effects of wife's death on the hazard of husband's death are limited in time, with considerable fading after about 2 months.

Although our working dependent variable is wife's death, the odds ratios must be interpreted as the effect of wife's death on the odds of husband's death. That's because of the time ordering of the observations—wife's death always precedes husband's death. If the goal were to estimate the effect of husband's death on wife's mortality, we would have to construct a completely different data set that would include couple-days prior to the wife's death, but not thereafter.

In this example, we estimated the effect of a single dichotomous covariate (wife's death within a specified number of days) on the occurrence of a nonrepeated event (husband's death). The method enabled us to control for all stable predictors. But suppose we want to control for time-varying predictors, such as smoking status. Simulation studies (Allison & Christakis, 2006) indicate that additional predictors can simply be included as predictor variables in the logistic regression model specified in Equation 5.6. Although the coefficients for any additional predictors would not be unbiased estimates of their effects on husband's death, the introduction of such predictors should yield approximately unbiased estimates for the effect of wife's death on husband's death (β in Equation 5.6). If we want to estimate the effect of, say, smoking status on husband's death, then we must make the probability of smoking be the dependent variable in Equation 5.6, possibly including wife's vital status as a covariate. This procedure could work even if smoking status had more than two categories, in which case Equation 5.6 would need to be specified as a multinomial logistic regression. However, I know of no way to generalize the case-time-control method to estimate the effects of quantitative predictors.

Summary

Fixed effects regression analysis of event history data typically requires that each individual has multiple, repeated, events. As we saw with logistic regression, the use of dummy variables to estimate the fixed effects usually leads to biased coefficient estimates for the other variables. This incidental parameters problem can be avoided by using Cox regression with stratification to eliminate the fixed effects from the partial likelihood, a method that is computationally efficient even for large numbers of strata. Under most conditions, the method of stratification produces approximately unbiased estimates.

As with other fixed effects methods, Cox regression with stratification may suffer a substantial loss of statistical power. Naturally, individuals with only one observation contribute nothing to the analysis. Even individuals with one censored and one uncensored interval are eliminated if the censored interval is the shorter of the two. Finally, only within-individual variation is used in estimating the coefficients. For reasons that are not fully understood, the hybrid method—which worked well for linear, logistic, and count data regression—does not produce correct results for Cox regression.

Serious difficulties arise in the attempt to do fixed effects regression analysis with nonrepeated events. The basic strategy is to treat time as discrete and create a separate record for each discrete time point that is observed for each individual, from the beginning of observation to the time of the event or censoring. For each record, a dichotomous dependent variable is coded 1 if an event occurred at that time point, otherwise 0. The final step is a conditional logistic regression of this dependent variable with stratification on individuals, using predictors that vary across time points. The fundamental problem with this appealing approach is that if time (or any monotonic function of time) is used as a predictor, the model will not converge due to separation. The reason is that the event always occurs at the end of each individual's sequence of records, so time perfectly predicts the occurrence of the event.

Although models that do not include time can certainly be estimated, the resulting coefficient estimates may be biased because effects of time on both the hazard and the predictors have not been controlled. One solution is the case-time-control method that appears to work well for estimating the effect of a categorical covariate on the hazard. The innovation of this method is to reverse the role of the dependent and independent variables in the conditional logistic regression, making it possible to include time as a covariate in the model.

Notes

1. I actually tried to do this, but my computer was still running after 10 days, at which point I terminated the job. In principle, such computational difficulties could be solved by using Greene's (2001) algorithms, but these are not currently available in commercial software.

2. I am grateful for Nicholas Christakis for permission to use these data, which are described in more detail in Allison and Christakis (2006).

3. It *is* possible to include non-monotonic functions of time such as $\sin(2\pi t/365)$, which would vary periodically over the course of a year.

4. This symmetry is exact when the model is "saturated" in the control predictors but only approximate for unsaturated models. A saturated model is one with only categorical predictors and all possible interactions.

CHAPTER 6. STRUCTURAL EQUATION MODELS WITH FIXED EFFECTS

In Chapter 2, we considered several different methods for estimating linear fixed effects regression models. In this chapter, we shall see how to estimate a fixed effects regression as a linear structural equation model with a latent variable. Why do we need another method for estimating the same model? The answer is that by putting the model into a structural equation framework we can accomplish several things that are difficult or impossible with conventional computational methods. In particular, we can

- estimate models that are a compromise between fixed and random effects,
- construct a likelihood ratio test for fixed versus random effects,
- estimate fixed effects models with reciprocal effects between the two response variables,
- estimate fixed effects models with lagged values of the response variable, and
- estimate models with multiple indicators of latent variables.

I've devoted a separate chapter to this method because the data structure and the conceptual framework are very different from that used for most of the methods described in Chapter 2. I first explain how to use structural equation software to estimate the random effects model described in Chapter 2. Then, we will see how that model can be modified to produce the fixed effects model.

Random Effects as a Latent Variable Model

In Chapter 2, the random effects model was specified as

$$y_{it} = \mu_t + \beta \mathbf{x}_{it} + \gamma \mathbf{z}_i + \alpha_i + \varepsilon_{it} \tag{6.1}$$

where y_{it} is the value of the response variable for individual i at time t, $\mathbf{x}_{it}$ is a vector of time-varying predictors, $\mathbf{z}_i$ is a vector of time-invariant predictors, α_i denotes the random effects, and ε_{it} is a random disturbance term. We assume that α_i and ε_{it} represent independent normally distributed variables with a mean of 0 and each having a constant variance. We also assume, at least for now, that these random components are independent of both $\mathbf{x}_{it}$ and $\mathbf{z}_i$.

It is well-known (Muthén, 1994) that random effects models such as those in Equation 6.1 can be represented as a structural equation model (SEM), which may be estimated with one of the many software programs that are designed for such models (e.g., LISREL, EQS, MX, Mplus, or Amos). Unfortunately, there is no Stata command for estimating such models.[1] Instead, I have used Mplus (www.statmodel.com) to estimate the models discussed in this chapter. Conceptually, we regard Equation 6.1 as specifying a separate equation for each point in time, with regression coefficients constrained to be the same across time periods. The random components, α and ε, are treated as latent variables. However, while there are separate epsilons for each time point, there is only one α that is common to all time periods.

SEMs are often represented as path diagrams (Kline, 2004). Figure 6.1 is a path diagram for a model with three time periods and a single time-varying independent variable. In path diagrams for SEMs, the convention is that directly observed variables are enclosed by rectangles while latent variables are enclosed by circles or ellipses. A straight, single-headed arrow denotes a direct causal effect of one variable on another, while a curved double-headed arrow denotes a bivariate correlation between two exogenous variables. (In the language of simultaneous equation models, endogenous variables are those that are dependent variables in at least one equation. Exogenous variables are those that are not dependent variables in any equation.)

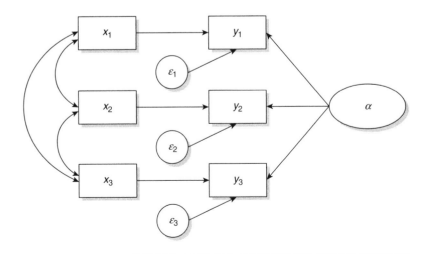

Figure 6.1   Path Diagram of a Random Effects Model for Three Periods

In Chapter 2, we estimated the model in Equation 6.1 using the Stata command **xtreg**, as applied to the NLSY data, which had observations at three periods for 581 children. The working data set had three records per child, for a total of 1,743 records. The dependent variable was a measure of antisocial behavior (ANTI). Independent variables included two time varying variables, poverty (POV) and self-esteem (SELF), along with several time-invariant variables.

To estimate the model as an SEM, we use the original form of the data set with one record per child and distinct variable names for the same variable measured at different times. Program code for accomplishing this with the Mplus package is shown in Appendix 2.

There are several points to keep in mind when writing SEM programs to estimate random effects models:

- The default in many SEM packages is to estimate the model using only information from the covariance matrix, in which case you will not get estimates of the intercepts, denoted by μ_t in Equation 6.1. If you want the intercepts, you must specify appropriate options that will incorporate the means into the analysis. This will not change the regression coefficients, however.
- The model is specified as three separate equations, for ANTI90, ANTI92, and ANTI94. The coefficients must be constrained to be equal across the three equations. In Mplus, this is accomplished by putting numbers in parentheses after the names of predictor variables and using the same numbers for parameters constrained to be the same. Relaxing these constraints is equivalent to allowing interactions between the predictor variables and time itself.
- One must also constrain the error variances of ε_1, ε_2, and ε_3 to be equal across the three equations.

As with most SEM programs, Mplus produces a large amount of output. A crucial part of this output—the regression coefficients, standard errors, and test statistics—is displayed in Table 6.1. These estimates should be compared with those in Table 2.5 produced by the **xtreg** command. The coefficients and standard errors are virtually identical.[2]

We now have a way of estimating a random effects model with SEM software that gives us the same results as the **xtreg** command in Stata. However, there are some important limitations to this method. First, unlike **xtreg**, this method may be hard to implement with unbalanced data. Data are balanced if the number of repeated measurements is the same for each individual in the sample. On the other hand, if some of the children in our sample had missing values for, say, ANTI94, some SEM software would

Table 6.1 SEMs for NLSY Data

	Random Effects		Fixed Effects		Compromise	
	Coefficient	Standard Error	Coefficient	Standard Error	Coefficient	Standard Error
SELF	−0.062**	0.009	−0.055**	0.011	−0.062**	0.009
POV	0.247**	0.080	0.112	0.093	0.111	0.093
BLACK	0.227	0.125	0.269*	0.126	0.269*	0.126
HISPANIC	−0.218	0.138	−0.198	0.138	−0.201	0.138
CHILDAGE	0.088	0.091	0.089	0.091	0.090	0.091
MARRIED	−0.049	0.126	−0.022	0.126	−0.025	0.126
GENDER	−0.483**	0.106	−0.476**	0.106	−0.479**	0.106
MOMAGE	−0.022	0.025	−0.026	0.025	−0.025	0.025
MOMWORK	0.261*	0.114	0.296**	0.115	0.295**	0.115

*.01 < p < .05.

**p < .01.

require that they be deleted entirely from the sample. Fortunately, Mplus and most other SEM packages now have options for maximum likelihood estimation with missing data that make it possible to handle such unbalanced data. Second, although possible, it's also cumbersome to set up the model to handle linear effects of time, linear interactions with time, or random coefficients (Muthén & Curran, 1997). In contrast, this is easily managed in **xtreg** and most random effects software.

Balancing these limitations are some important advantages to the SEM approach. First, it is possible to combine the random effects model with models for multiple indicators of latent variables. These variables may be either independent or dependent variables. Good introductions to multiple-indicator latent variable models can be found in Kline (2004) or Hatcher (1994). Second, as we will see in the next section, the random effects model in the SEM framework can be extended to estimate fixed effects models in ways that facilitate a comparison and a compromise between the two models.

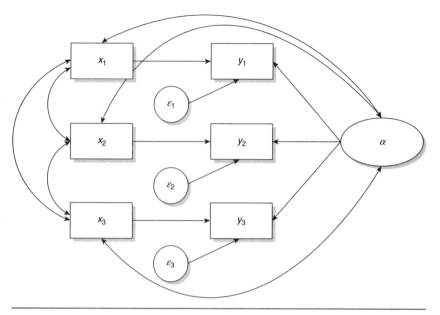

Figure 6.2 Path Diagram for a Fixed Effects Model for Three Periods

Fixed Effects as a Latent Variable Model

As previously noted in Chapter 2, the basic random effects model is actually a special case of the fixed effects model (Mundlak, 1978). The random effects model assumes that α_i is uncorrelated with $\mathbf{x}_{it}$, the vector of time-varying predictors. The fixed effects model allows for *any* correlations between α_i and the elements of $\mathbf{x}_{it}$. Figure 6.2 shows a path diagram for a simplified fixed effects model with a single, time-varying predictor. The only difference between this diagram and the one in Figure 6.1 is the addition of the curved arrows representing correlations between α and the x variables.

These additional correlations can be easily incorporated into SEM software by simply specifying the correlations between the latent variable and the time-varying predictors (Allison & Bollen, 1997; Teachman, Duncan, Yeung, & Levy, 2001). Note that a correlation cannot be allowed between the latent variable and any time-invariant predictors such as GENDER or MARRIED. Attempting to do so results in something called an underidentified model, which will typically produce an error or warning message.

The coefficient estimates and associated statistics for the fixed effects model are shown in the second panel of Table 6.1. Looking first at the coefficients and standard errors for SELF and POV, we see that they are identical to the estimates in Table 2.5, estimated with the fixed effects option in **xtreg**.[3] They are also identical to the results in Table 2.8 obtained using the hybrid method.

Like the hybrid method, Table 6.1 also gives coefficient estimates for the time-invariant variables. However, the estimates and test statistics in Table 6.1 for these variables are quite different from the estimates and test statistics in Table 2.8. For example, the coefficient for MOMWORK in Table 6.1 is clearly statistically significant, but is just as clearly not significant in Table 2.8. Which is better, the hybrid estimates or the SEM estimates? Well, it depends. Simulation results (not shown) strongly suggest that when the correlations between the time-invariant covariates z and the unobserved heterogeneity term α are zero, the estimates produced by SEM are approximately unbiased but the hybrid estimates may be substantially biased. On the other hand, when z and α are correlated, both estimates are biased but the SEM estimates show more bias than the hybrid estimates.

Now that we have both a fixed and random effects version of our SEM, it is a simple matter to produce a likelihood-ratio statistic to compare them. For each model, the output contains a chi-square statistic and associated degrees of freedom. This statistic compares the overall fit of the model to a saturated model that perfectly reproduces the covariance matrix for all the variables. For the random effects model, the chi-square is 84.42 with 34 degrees of freedom. For the fixed effects model, it's 66.45 with 28 degrees of freedom. The difference between the two is a chi-square of 17.97 and 6 degrees of freedom. The 6 degrees of freedom correspond to the six additional correlations that are allowed under the fixed effects model. The p value for this chi-square is .006, indicating that we should reject the random effects model in favor of the fixed effects model. This is the same conclusion that we reached in Chapter 2 using either the Hausman test produced by Stata or the tests of equality for coefficients of the mean and centered scores.[4] Like the hybrid test, the likelihood ratio test computed here may have better statistical properties than the Hausman test which could, for example, have negative values for some data configurations.

A Compromise Between
Fixed Effects and Random Effects

In the last section, we obtained a fixed effects model by starting with a random effects model and then allowing for all possible correlations between the random effect α and the time-varying explanatory variables. But perhaps all those correlations aren't really needed. Table 6.2 shows estimated correlations and covariances between α and the time-varying variables produced

Table 6.2 Correlations Between α and Time-Varying Predictors

	Correlation	*z Statistic*
SELF90	−.006	−0.77
SELF92	−.0146	−1.71
SELF94	−.008	−1.01
POV90	.123	3.34
POV92	.049	1.33
POV94	.095	2.49

by Mplus. It appears that the correlations with the SELF variables are very small and not statistically significant, while the correlations with the POV variables are somewhat larger and two of the three are statistically significant. This suggests that we could set the SELF correlations equal to zero without appreciably worsening the fit of the model. That's desirable because the estimate for SELF would then be based on both within- and between-person variation, yielding a smaller standard error.[5]

This is easily accomplished in Mplus and produces the results shown in the third panel of Table 6.1. The coefficient and t statistic for POV is about the same as what we found for the fixed effects model. On the other hand, the coefficient for SELF is somewhat larger than in the pure fixed effects model, and its standard error is about 20% smaller. Taking the difference in the chi-squares for the two models, we get a chi-square of 3.00 with 3 degrees of freedom. This is definitely not statistically significant, indicating that we cannot reject the simpler model (which set three correlations equal to 0) in favor of the more complicated model.

Reciprocal Effects With Lagged Predictors

We have just seen that many of the fixed and random effects models estimated in Chapter 2 can also be estimated with SEM software, and that there are both advantages and disadvantages to this approach. We are now going to consider some important fixed effects models that go considerably beyond those in Chapter 2, and which can be conveniently estimated in a structural equation framework. These models violate the strict exogeneity assumption of Chapter 2, which stated that $\mathbf{x}_{it}$ is statistically independent of $\varepsilon_{it'}$ for any t and t'. This happens either because $\mathbf{x}_{it}$ is affected by y at an earlier point in time, or because one component of $\mathbf{x}_{it}$ is y itself at an earlier point in time (a lagged dependent variable). These models are important because they offer the possibility of enhancing our ability to determine the direction of causality among variables that are associated with one another.

Let's suppose that we observe two variables, x and y, that are known to be correlated. We would like to know whether x causes y or y causes x (or perhaps both). Both variables are observed at several points in time. Consider the following model:

$$y_{it} = \mu_t + \beta x_{i(t-1)} + \alpha_i + \varepsilon_{it}$$

$$x_{it} = \tau_t + \delta y_{i(t-1)} + \eta_i + \upsilon_{it} \tag{6.2}$$

This model says that y is affected by x at an earlier time point and x is affected by y at an earlier time point. The model also includes fixed effects α and η, representing the effects of any and all time-invariant predictors on each variable. We could also include other lagged time-varying predictors as well as time-invariant predictors, but that would unnecessarily complicate the discussion.

How can this model be estimated? If there are observations at exactly three time points, the model can be estimated by taking first differences and applying ordinary least squares (OLS) to each equation separately:[6]

$$y_{i3} - y_{i2} = (\mu_3 - \mu_2) + \beta(x_{i2} - x_{i1}) + (\varepsilon_{i3} - \varepsilon_{i2})$$

$$x_{i3} - x_{i2} = (\tau_3 - \tau_2) + \delta(y_{i2} - y_{i1}) + (\upsilon_{i3} - \upsilon_{i2}) \tag{6.3}$$

When there are more than three time points, it might seem that the methods used in Chapter 2 (dummy variables for individuals or deviations from the means) might do the job. Unfortunately, because there are reciprocal effects, the deviation scores used in fixed effects estimation are necessarily correlated with the error terms in the regressions, and that leads to biased estimation (Wooldridge, 2002). Fortunately, using the method for incorporating fixed effects into an SEM, we can circumvent these difficulties.

Even more serious difficulties arise when the model is further extended to allow for lagged values of the dependent (endogenous) variables:

$$y_{it} = \mu_t + \beta_1 x_{i(t-1)} + \beta_2 y_{i(t-1)} + \alpha_i + \varepsilon_{it}$$

$$x_{it} = \tau_t + \delta_1 x_{i(t-1)} + \delta_2 y_{i(t-1)} + \eta_i + \upsilon_{it} \tag{6.4}$$

If we exclude the fixed effects (α and η), this model is well-known in the social science literature as the two-wave, two-variable panel model or the cross-lagged panel model.

In the econometric literature, panel models with lagged dependent variables are referred to as *dynamic* models. They are well-known to pose serious difficulties for conventional estimation methods, and several alternative methods have been proposed to deal with them (Baltagi, 1995; Honoré, 1993; Honoré & Kyriazidou, 2000). The methods generally rely on the use of lagged variables in an instrumental variables (IV) framework. The best-known method is that proposed by Arellano and Bond (1991), which has been implemented in the Stata command **xtabond**. Lancaster (2000), however, has described the IV approach as "odd," and used "only because econometricians do not know how to use the likelihood correctly!"

It turns out that maximum likelihood estimates of dynamic, fixed effects models can be estimated in a straightforward way using SEM software. Although the properties of this method have not been investigated analytically, my own simulation studies (Allison, 2000) have shown that it does an excellent job of recovering the parameters for models such as Equation 6.4.

As an example, I analyzed data for 178 occupations in the United States for the years 1983, 1989, 1995, and 2001. The data come from the March "Current Population Survey: Annual Demographic File" (CPS). The observations in the original CPS data are individual people, but I used only aggregated data for the 178 occupations. For occupation in each year, I calculated the proportion female and the median wage for females. This was done only for the 178 occupations that had at least 50 sample members in each of the years. Further details can be found in England, Allison, and Wu (2007). For wages, the variables are labeled MDWGF1-MDWGF4, and for proportion female we have PF1-PF4.

For the model in Equation 6.4, let y be median wage and let x be proportion female. In 1983, the correlation between these two variables was -0.33, which was highly significant. There has been considerable controversy regarding the possible direction of causality between these two variables (England et al., 2007). One argument is that employers devalue occupations that have a high proportion female and, consequently, pay lower wages. The rival hypothesis is that declining wages make occupations less attractive to men; as they leave for better-paying work, women fill their vacant positions. I shall assume that changes in either of these variables show up in changes in the other variable 6 years later.

By estimating the two equations in Equation 6.4, we can assess each of the two possible causal effects. Although it's possible to estimate the two equations simultaneously, estimating them separately allows for considerably more flexibility in specifying the model.[7] In addition to the fixed effects, the key device that allows for the reciprocal effects is this: The error term at each point in time must be allowed to correlate with *future* values of the time-dependent covariate (Wooldridge, 2002). In our example, the error term in the equation for median wage at Time 2 must be allowed to have a nonzero correlation with percentage female at Time 3. Similarly, the error term in the equation for percent female at Time 2 must be correlated with median wage at Time 3. Note that there is no equation predicting median wage or proportion female at Time 1 because we do not observe their lagged values 6 years earlier (1977).

Note also that for the lagged dependent variable, a correlation is only allowed between the latent variable and the value of the variable at Time 1. That's because only the Time 1 variable is exogenous and correlations are only allowed among exogenous variables. There's actually no need to specify a correlation between the latent variable and the later values of the lagged dependent variable because the latent variable is one of the predictors in the equation for each of these variables.

Results for the two equations are shown in Table 6.3. Not surprisingly, each variable has a positive, statistically significant effect on itself 6 years later. With respect to the "cross-lagged" coefficients, however, there is no evidence for an effect in either direction.

Elsewhere, I have questioned the desirability of including lagged values of the dependent variable as a predictor when fixed effects are already in the model (Allison, 1990). So I also estimated a model that removes the lagged dependent variables, and got essentially the same

Table 6.3 Estimates for Reciprocal Effects Model

	Response Variable			
	Median Wage		Proportion Female	
Predictor	Coefficient	Standard Error	Coefficient	Standard Error
Median wage	0.344**	0.064	−0.001	0.002
Proportion female	−0.159	2.447	0.299**	0.079

*.01 < p < .05.

**p < .01.

results for the cross-lagged coefficients. Similarly, a model that includes the lagged dependent variables but does *not* include the fixed effects (the classic two-wave, two-variable panel model) yields no evidence for a cross-lagged effect in either direction.

Summary

Linear, fixed effects or random effects regression models for quantitative response variables can be estimated with SEM software to yield the same results as those obtained using the more conventional methods described in Chapter 2. This approach requires a different data structure, however, with one record containing all the measurements for each individual or cluster, and with the multiple measurements coded as distinct variables. In SEM software, a separate equation is specified for each response variable at each point in time, and the coefficients are typically constrained to be the same across equations. The random or fixed effect is specified as a latent variable that is common to all the equations. In the fixed effects version, this latent variable is allowed to be correlated with all the predictor variables that vary across equations.

This approach is typically more cumbersome to set up than the methodology described in Chapter 2. Nevertheless, it allows for a number of interesting extensions, including a likelihood ratio test comparing fixed and random effects, a compromise between fixed and random effects models, and models that have multiple indicators of latent variables. Most important, within the SEM framework it is possible to estimate models for panel data in which two or more response variables are believed to have lagged, reciprocal effects on each other. Such models allow for much stronger causal inferences from nonexperimental data than is ordinarily the case.

Notes

1. It is possible to estimate some SEMs in Stata using the user-contributed command **gllamm**. But even with this command, the setup is rather awkward and complex.

2. I have not reported coefficients for TIME in Table 6.1 because the three intercepts estimated by Mplus do not exactly correspond to the coefficients in Table 2.5. The TIME2 coefficient in Table 2.5 is equal to the difference between the intercept for Time 2 and the intercept for Time 1. Similarly, the TIME3 coefficient in Table 2.5 is equal to the difference between the intercept for Time 3 and the intercept for Time 1.

3. Ejrnaes and Holm (2006) incorrectly claim that the traditional fixed effects estimator is distinct from the SEM estimator. In fact, the two methods always give identical results.

4. The SEM test has 6 degrees of freedom, the Hausman test had 4, and the hybrid test had only 2. That is because the SEM test allows the covariances between α and x to be different at each of the three time periods while the other two implicitly constrain them to be the same. The Hausman test has 2 more df than the hybrid test because it also tests whether the two time coefficients are the same in the random effects and fixed effects models. We can get a 2 df test in the SEM framework by constraining the covariances between α and x to be the same for each period when fitting the fixed effects model. For the NLSY example, that yields a chi-square value of 11.41 with 2 df, for a p value of .003, slightly smaller than the p value for the 6 df test. That's quite similar to the hybrid test which produced a chi-square of 9.86 with 2 df, for a p value of .007.

5. See Hausman and Taylor (1981) for a different approach using IV.

6. The reason these equations can be estimated by OLS is that both xs in the first equation are independent of both εs. And that is because xs are only affected (indirectly) by ε at earlier points in time. The same rationale applies to the second equation.

7. If the two equations are estimated simultaneously, each variable must be expressed in the same way both as a dependent and as an independent variable. If they are estimated separately, on the other hand, one could use, for example, the logarithm of the variable as a dependent variable and the unlogged version as an independent variable.

APPENDIX 1. STATA PROGRAMS
FOR EXAMPLES IN CHAPTERS 2 TO 5

```
use "c:\data\nlsy", clear

/* Table 2.1 */
reg   anti90 self90 pov90
reg   anti94 self94 pov94
gen antidiff = anti94 - anti90
gen povdiff = pov94 - pov90
gen selfdiff = self94 - self90
reg antidiff povdiff selfdiff

/* Table 2.2 */
reg antidiff povdiff selfdiff pov90 self90 black ///
  hispanic childage married gender momage momwork

/* Table 2.3 */
gen antidif1=anti92-anti90
gen antidif2=anti94-anti92
gen selfdif1=self92-self90
gen selfdif2=self94-self92
gen povdif1=pov92-pov90
gen povdif2=pov94-pov92
reg antidif2 selfdif2 povdif2
reg antidif1 selfdif1 povdif1
gen id=_n
reshape long antidif povdif selfdif, i(id)
gen eqdum=_j-1
reg antidif povdif selfdif
xtset id _j
xtreg antidif povdif selfdif eqdum, pa
```

100

```
/* create data set with 3 records per person */
use "c:\data\nlsy", clear
gen id=_n
reshape long anti self pov, i(id)
gen time=1+(_j-90)/2

/* Table 2.5 */
xi: reg anti self pov i.time i.id
xi: reg anti self pov i.time
xtset id time
xi: xtreg anti self pov i.time, fe

/* Table 2.6 */
xi: xtreg anti i.time*self i.time*pov ///
   i.time*gender i.time*childage ///
   i.time*hispanic i.time*black i.time*momwork ///
   i.time*married ///
   i.time*momage, fe i(id)
testparm _ItimXself* _ItimXpov* _ItimXgend* ///
   _ItimXchil* _ItimXhisp* ///
   _ItimXblac* _ItimXmomw* _ItimXmarr* _ItimXmoma*

/* Table 2.7 */
xi: xtreg anti self pov i.time black hispanic ///
   childage married gender momage momwork
xi: xtreg anti self pov i.time


/* Hausman test */
xi: xtreg anti self pov gender childage hispanic ///
   black momwork married momage i.time
estimates store random_effects
xi: xtreg anti self pov i.time, fe
estimates store fixed_effects
hausman fixed_effects random_effects

/* Table 2.8 */
egen mself=mean(self), by(id)
egen mpov=mean(pov), by(id)
gen dself=self-mself
gen dpov=pov-mpov
```

```
xi: xtreg anti dself dpov mself mpov black ///
   hispanic childage married ///
   gender momage momwork i.time
test (dself=mself) (dpov=mpov)
xi: xtmixed anti dself dpov mself mpov black ///
   hispanic childage married ///
   gender momage momwork i.time ||id: dself


/*Table 3.1*/
use "c:\data\teenpov.dta", clear
tab pov1 pov5


/* Table 3.2 */
drop if pov1==pov5
gen dmother=mother5-mother1
gen dspouse=spouse5-spouse1
gen dschool=inschool5-inschool1
gen dhours=hours5-hours1
logit pov5 dmother dspouse dschool dhours
logit pov5 dmother dspouse dschool dhours black age
logit pov5 dmother dspouse dschool dhours black ///
   age mother1 spouse1 inschool1 hours1


/* Table 3.4 */
use "c:\data\teenpov.dta", clear
reshape long pov mother spouse inschool hours, ///
   i(id)
rename inschool school
rename _j year
xtset id year
xi: xtlogit pov mother spouse school hours ///
   i.year, fe
xi: xtlogit pov mother spouse school hours ///
   i.year, pa corr(uns)
xi: xtlogit pov mother spouse school hours i.year


/* Table 3.5 */
gen mothblack=mother*black
xi: xtlogit pov mother spouse school hours ///
   mothblack i.year, fe
gen yearschool=(year-1)*school
```

```
gen yearhours=(year-1)*(hours-8.67)
gen yearblack=(year-1)*black
gen yearage=(year-1)*(age-15.65)
xi: xtlogit pov mother spouse school hours year ///
  yearschool yearhours ///
  yearblack yearage, fe

/* Table 3.6, 3.7 */
egen mmother=mean(mother), by(id)
egen mspouse=mean(spouse), by(id)
egen mschool=mean(school), by(id)
egen mhours=mean(hours), by(id)
gen dmother=mother-mmother
gen dspouse=spouse-mspouse
gen dschool=school-mschool
gen dhours=hours-mhours
xi:  xtlogit pov dmother dspouse dschool dhours ///
  mmother mspouse ///
  mschool mhours black age i.year
test dmother=mmother
test dspouse=mspouse
test dschool=mschool
test dhours=mhours
test (dmother=mmother) (dspouse=mspouse) ///
  (dschool=mschool) (dhours=mhours)
xi:  xtmelogit pov dmother dspouse dschool dhours ///
  mmother mspouse ///
  mschool mhours black age i.year ||id: dmother

/* Table 3.8 */
use "c:\data\nlsy", clear
gen id=_n
reshape long anti self pov, i(id)
gen time=1+(_j-90)/2
egen mself=mean(self), by(id)
egen mpov=mean(pov), by(id)
gen dself=self-mself
gen dpov=pov-mpov
xi: ologit anti dself dpov mself mpov black ///
  hispanic childage married ///
  gender momage momwork i.time, cluster(id)
test (dself=mself) (dpov=mpov)
```

```
/* Table 3.9 */
use "c:\data\teenpov2.dta", clear
reshape long mother spouse empstat, i(id)
drop if empstat==.
gen currage=age+_j-1
egen mmother=mean(mother), by(id)
egen mspouse=mean(spouse), by(id)
egen mage=mean(currage), by (id)
gen dmother=mother-mmother
gen dspouse=spouse-mspouse
gen dage=currage-mage
mlogit empstat dmother mmother dspouse mspouse ///
  dage mage black, ///
  vce(cluster id) base(1)
test ([ #1] dmother=[ #1] mmother) ///
  ([ #1] dspouse=[ #1] mspouse) ([ #1] dage=[ #1] mage)
test ([ #2] dmother=[ #2] mmother) ///
  ([ #2] dspouse=[ #2] mspouse) ([ #2] dage=[ #2] mage)
preserve
drop if empstat==3
gen empstat2=empstat-1
xtset id _j
xtlogit empstat2 dmother ///
  mmother dspouse mspouse dage mage black, re
drop if empstat==2
gen empstat3=empstat-1
xtset id _j
xtlogit empstat3 dmother mmother dspouse mspouse ///
  dage mage black, re

/* Table 4.1 */
use patents, clear
gen total=pat75+pat79
gen   rd0=logr79-logr75
gen   rd1=logr78-logr74
gen   rd2=logr77-logr73
gen   rd3=logr76-logr72
gen   rd4=logr75-logr71
gen   rd5=logr74-logr70

blogit pat79 total
blogit pat79 total, vce(jack)
```

```
blogit pat79 total, vce(boot)
blogit pat79 total rd0-rd5
blogit pat79 total rd0-rd5, vce(jack)
blogit pat79 total rd0-rd5, vce(boot)

blogit pat79 total rd0-rd5 science logsize
blogit pat79 total rd0-rd5 science logsize, ///
  vce(boot)

/* Table 4.2 */
use patents, clear
rename pat75 patent1
rename pat76 patent2
rename pat77 patent3
rename pat78 patent4
rename pat79 patent5
gen sumpat=patent1+patent2+patent3+patent4+patent5
gen rda1=logr75
gen rda2=logr76
gen rda3=logr77
gen rda4=logr78
gen rda5=logr79
gen rdb1=logr74
gen rdb2=logr75
gen rdb3=logr76
gen rdb4=logr77
gen rdb5=logr78
gen rdc1=logr73
gen rdc2=logr74
gen rdc3=logr75
gen rdc4=logr76
gen rdc5=logr77
gen rdd1=logr72
gen rdd2=logr73
gen rdd3=logr74
gen rdd4=logr75
gen rdd5=logr76
gen rde1=logr71
gen rde2=logr72
gen rde3=logr73
gen rde4=logr74
gen rde5=logr75
```

```
gen  rdf1=logr70
gen  rdf2=logr71
gen  rdf3=logr72
gen  rdf4=logr73
gen  rdf5=logr74
gen  id=_n
reshape long patent rda rdb rdc rdd rde rdf, i(id)
rename  _j time
rename  rda rd0
rename  rdb rd1
rename  rdc rd2
rename  rdd rd3
rename  rde rd4
rename  rdf rd5
list id time patent rd0-rd5 in 1/20
xtset id time

/* Table 4.3 */
xi: xtpoisson patent rd0 rd1 rd2 rd3 rd4 rd5 ///
   i.time, fe
xi: xtpoisson patent rd0 rd1 rd2 rd3 rd4 rd5 ///
   i.time, fe vce(boot)
xi: xtpoisson patent rd0 rd1 rd2 rd3 rd4 rd5 ///
   i.time, re
xi: xtpoisson patent rd0 rd1 rd2 rd3 rd4 rd5 ///
   i.time, re vce(boot)
xi: xtpoisson patent rd0 rd1 rd2 rd3 rd4 rd5 ///
   i.time, pa corr(uns) vce(robust)
xi: poisson patent i.id rd0 rd1 rd2 rd3 rd4 rd5 ///
   i.time

/* Table 4.4 */
gen rdsci=rd0*science
xi: xtpoisson patent rd0 rdsci i.time, fe
xi: xtpoisson patent rd0 rdsci i.time, fe vce(boot)

/* Table 4.5 */
gen scitime=time*science
xtpoisson patent rd0 time scitime, fe i(id)
xtpoisson patent rd0 time scitime, fe i(id) ///
  vce(boot)
```

```
/* Table 4.6 */
drop if sumpat==0
xi: nbreg patent i.id rd0 rd1 rd2 rd3 rd4 rd5 ///
  i.time
xi: nbreg patent i.id rd0-rd5 i.time, vce(opg)

/* Table 4.7 */
xi: xtnbreg patent rd0 rd1 rd2 rd3 rd4 rd5 ///
  i.time, fe i(id)
xi: xtnbreg patent rd0 rd1 rd2 rd3 rd4 rd5 ///
  science logsize i.time, fe i(id)

/* Table 4.8 */
egen mrd0=mean(rd0), by(id)
egen mrd1=mean(rd1), by(id)
egen mrd2=mean(rd2), by(id)
egen mrd3=mean(rd3), by(id)
egen mrd4=mean(rd4), by(id)
egen mrd5=mean(rd5), by(id)
gen drd0=rd0-mrd0
gen drd1=rd1-mrd1
gen drd2=rd2-mrd2
gen drd3=rd3-mrd3
gen drd4=rd4-mrd4
gen drd5=rd5-mrd5
xi: xtnbreg patent drd0 drd1 drd2 drd3 drd4 drd5 ///
  mrd0 mrd1 mrd2 mrd3 ///
  mrd4 mrd5 science logsize i.time, re
test (drd0=mrd0)(drd1=mrd1)(drd2=mrd2) ///
(drd3=mrd3)(drd4=mrd4)(drd5=mrd5)
xi: xtnbreg patent drd0 drd1 drd2 drd3 drd4 drd5 ///
  mrd0 mrd1 mrd2 mrd3 ///
  mrd4 mrd5 science logsize i.time, pa robust
test (drd0=mrd0)(drd1=mrd1)(drd2=mrd2) ///
  (drd3=mrd3)(drd4=mrd4)(drd5=mrd5)

/* Table 5.1 */
use "C:\data\nsfg.dta", clear
stset dur, failure(birth==1)
stcox pregordr age married passt nobreast lbw ///
caesar multiple college, nohr
```

```
stcox pregordr age married passt nobreast lbw ///
  caesar multiple college, ///
  nohr cluster(caseid)

/* Table 5.2 */
stcox pregordr age married passt nobreast lbw ///
  caesar multiple college, ///
  strata(caseid) nohr
gen collbreast=college*nobreast
stcox pregordr age married passt nobreast lbw ///
  caesar multiple college ///
    collbreast, nohr strata(caseid)

/* Table 5.3 */
streg pregordr age married passt nobreast lbw ///
  caesar multiple college, ///
  d(gompertz) nohr shared(caseid)

/* Table 5.4 */
use "C:\data\coupleday.dta", clear
xtset coupleid day
xtlogit husdead wifed15, fe or
xtlogit husdead wifed30, fe or
xtlogit husdead wifed60, fe or
xtlogit husdead wifed90, fe or
xtlogit husdead wifed120, fe or
logit husdead wifed15, or
logit husdead wifed30, or
logit husdead wifed60, or
logit husdead wifed90, or
logit husdead wifed120, or

/* Table 5.5 */
drop if wifefirst==0
gen day2=day*day
xtlogit wifed15 husdead day day2, fe or
xtlogit wifed30 husdead day day2, fe or
xtlogit wifed60 husdead day day2, fe or
xtlogit wifed90 husdead day day2, fe or
xtlogit wifed120 husdead day day2, fe or
```

APPENDIX 2. MPLUS PROGRAMS FOR EXAMPLES IN CHAPTER 6

```
! Table 6.1
! Random Effects
Data: file is "c:\data\nlsy.dat";
Variable: names are anti90 anti92 anti94 black
childage gender hispanic married momage momwork
pov90 pov92 pov94 self90 self92 self94; usevariables =
anti90 anti92 anti94 black childage gender hispanic
married momage momwork pov90 pov92 pov94 self90
self92 self94 ;
Model:
  falpha by anti90-anti94@1;
  anti90 on
    pov90 (1)
    self90 (2)
    black (3)
    hispanic (4)
    childage (5)
    married (6)
    gender (7)
    momage (8)
    momwork (9);
  anti92 on
    pov92 (1)
    self92 (2)
    black (3)
    hispanic (4)
    childage (5)
    married (6)
    gender (7)
    momage (8)
    momwork (9);
```

```
anti94 on
    pov94 (1)
    self94 (2)
    black (3)
    hispanic (4)
    childage (5)
    married (6)
    gender (7)
    momage (8)
    momwork (9);
falpha with pov90-pov94@0 self90-self94@0 black@0
hispanic@0 childage@0 married@0 gender@0 momage@0
momwork@0;
anti90 anti92 anti94 (10);

! Fixed Effects
Data: file is "c:\data\nlsy.dat";
Variable: names are anti90 anti92 anti94 black
childage gender hispanic married momage momwork
pov90   pov92   pov94   self90   self92   self94;
usevariables  are  anti90  anti92  anti94  black
childage gender hispanic married momage momwork
pov90 pov92 pov94 self90 self92 self94 ;
Model:
  falpha by anti90-anti94@1;
  anti90 on
    pov90 (1)
    self90 (2)
    black (3)
    hispanic (4)
    childage (5)
    married (6)
    gender (7)
    momage (8)
    momwork (9);
  anti92 on
    pov92 (1)
    self92 (2)
    black (3)
    hispanic (4)
```

```
    childage (5)
    married (6)
    gender (7)
    momage (8)
    momwork (9);
anti94 on
    pov94 (1)
    self94 (2)
    black (3)
    hispanic (4)
    childage (5)
    married (6)
    gender (7)
    momage (8)
    momwork (9);
  falpha with black@0 hispanic@0 childage@0
    married@0 gender@0 momage@0 momwork@0;
  anti90 anti92 anti94 (10);

! Compromise
Data: file is "c:\data\nlsy.dat";
Variable: names are anti90 anti92 anti94 black
childage gender hispanic married momage momwork
pov90 pov92 pov94 self90 self92 self94; usevariables =
anti90 anti92 anti94 black childage gender hispanic
married momage momwork pov90 pov92 pov94 self90
self92 self94 ;
Model:
  falpha by anti90-anti94@1;
  anti90 on
    pov90 (1)
    self90 (2)
    black (3)
    hispanic (4)
    childage (5)
    married (6)
    gender (7)
    momage (8)
    momwork (9);
  anti92 on
    pov92 (1)
    self92 (2)
```

```
      black (3)
      hispanic (4)
      childage (5)
      married (6)
      gender (7)
      momage (8)
      momwork (9);
anti94 on
      pov94 (1)
      self94 (2)
      black (3)
      hispanic (4)
      childage (5)
      married (6)
      gender (7)
      momage (8)
      momwork (9);
    falpha with self90-self94@0 black@0 hispanic@0
childage@0
        married@0 gender@0 momage@0 momwork@0;
    anti90 anti92 anti94 (10);

! Table 6.3
Data: file is "c:\data\occ.dat";
Variable: names are pf1-pf4 mdwgf1-mdwgf4;
 usevariables pf1-pf4 mdwgf1-mdwgf3;
Model:
   alpha by pf2-pf4@1;
   pf4 on
     pf3 (1)
     mdwgf3 (2);
   pf3 on
     pf2 (1)
     mdwgf2 (2);
   pf2 on
     pf1 (1)
     mdwgf1 (2);
   mdwgf3 with pf2;

Data: file is "c:\data\occ.dat";
Variable: names are pf1-pf4 mdwgf1-mdwgf4;
 usevariables pf1-pf3 mdwgf1-mdwgf4;
```

```
Model:
  alpha by mdwgf2-mdwgf4@1;
  mdwgf4 on
    pf3 (1)
    mdwgf3 (2);
  mdwgf3 on
    pf2 (1)
    mdwgf2 (2);
  mdwgf2 on
    pf1 (1)
    mdwgf1 (2);
  mdwgf2 with pf3;
```

REFERENCES

Abrevaya, J. (1997). The equivalence of two estimators of the fixed-effects logit model. *Economics Letters, 55,* 41–44.

Albert, A., & Anderson, J. A. (1984). On the existence of maximum likelihood estimates in logistic regression models. *Biometrika, 71,* 1–10.

Allison, P. D. (1984). *Event history analysis: Regression for longitudinal event data.* Beverly Hills, CA: Sage.

Allison, P. D. (1990). Change scores as dependent variables in regression analysis. In C. Clogg (Ed.), *Sociological methodology 1990* (pp. 93–114). Oxford, UK: Basil Blackwell.

Allison, P. D. (1995). *Survival analysis using SAS.* Cary, NC: SAS Institute.

Allison, P. D. (1996). Fixed effects partial likelihood for repeated events. *Sociological Methods & Research, 25,* 207–222.

Allison, P. D. (1999a). *Logistic regression using SAS: Theory and application.* Cary, NC: SAS Institute.

Allison, P. D. (1999b). *Multiple regression: A primer.* Thousand Oaks, CA: Pine Forge.

Allison, P. D. (2000, June). *Inferring causal order from panel data.* Paper prepared for presentation at the Ninth International Conference on Panel Data, Geneva, Switzerland.

Allison, P. D. (2002). *Bias in fixed-effects Cox regression with dummy variables.* Unpublished paper, Department of Sociology, University of Pennsylvania.

Allison, P. D. (2004). Convergence problems in logistic regression. In M. Altman, J. Gill, & M. McDonald (Eds.), *Numerical issues in statistical computing for the social scientist* (pp. 247–262). New York: Wiley InterScience.

Allison, P. D. (2005). *Fixed effects regression methods for longitudinal data using SAS.* Cary, NC: SAS Institute.

Allison, P. D., & Bollen, K. A. (1997, August). *Change score, fixed effects, and random component models: A structural equation approach.* Paper presented at the annual meeting of the American Sociological Association.

Allison, P. D., & Christakis, N. (2006). Fixed effects methods for the analysis of non-repeated events. In R. Stolzenberg (Ed.), *Sociological methodology 2006* (pp. 155–172). Oxford, UK: Basil Blackwell.

Allison, P. D., & Waterman, R. (2002). Fixed effects negative binomial regression models. In R. M. Stolzenberg (Ed.), *Sociological methodology 2002* (pp. 247–265). Oxford, UK: Basil Blackwell.

Arellano, M., & Bond, S. (1991). Some tests of specification for panel data: Monte Carlo evidence and an application to employment equations. *Review of Economic Studies, 58,* 277–297.

Baltagi, B. H. (1995). *Econometric analysis of panel data.* New York: Wiley.

Begg, C. B., & Gray, R. (1984). Calculation of polychotomous logistic regression parameters using individualized regressions. *Biometrika, 71,* 11–18.

Bryk, A. S., & Raudenbusch, S. W. (1992). *Hierarchical linear models: Application and data analysis methods.* Newbury Park, CA: Sage.

Cameron, A. C., & Trivedi, P. K. (1998). *Regression analysis of count data.* Cambridge, UK: Cambridge University Press.

Center for Human Resource Research. (2002). *NLSY97 user's guide.* Washington, DC: U.S. Department of Labor.

Chamberlain, G. A. (1980). Analysis of covariance with qualitative data. *The Review of Economic Studies, 47,* 225–238.

Chamberlain, G. A. (1985). Heterogeneity, omitted variable bias, and duration dependence. In J. J. Heckman & B. Singer (Eds.), *Longitudinal analysis of labor market data* (pp. 3–38). Cambridge, UK: Cambridge University Press.

Conaway, M. R. (1989). Analysis of repeated categorical measurements with conditional likelihood methods. *Journal of the American Statistical Association, 84,* 53–62.

Cox, D. R. (1972). Regression models and life tables (with discussion). *Journal of the Royal Statistical Society, Series B, 34,* 187–220.

Darroch, J. N., & McCloud, P. I. (1986). Category distinguishability and observer agreement. *Australian Journal of Statistics, 28,* 371–388.

Dunteman, G. H., & Ho, M. R. (2005). *An introduction to generalized linear models.* Thousand Oaks, CA: Sage.

Ejrnaes, M., & Holm, A. (2006). Comparing fixed effects and covariance structure estimators for panel data. *Sociological Methods & Research, 35,* 61–83.

England, P., Allison, P. D., & Wu, Y. (2007). Does bad pay cause occupations to feminize, does feminization reduce pay, and how can we tell with longitudinal data? *Social Science Research, 36*(3), 1237–1256.

Goldstein, H. (1987). *Multilevel models in educational and social research.* London: Griffin.

Greene, W. H. (2000). *Econometric analysis* (4th ed.). Upper Saddle River, NJ: Prentice Hall.

Greene, W. H. (2001). *Estimating econometric models with fixed effects.* New York University, Leonard N. Stern School, Finance Department Working Paper Series.

Greenland, S. (1996). Confounding and exposure trends in case-crossover and case-time control designs. *Epidemiology, 7,* 231–239.

Hall, B. H., Griliches, Z., & Hausman, J. A. (1986). Patents and R and D: Is there a lag? *International Economic Review, 27*(2), 265–283.

Hatcher, L. (1994). *A step-by-step approach to using the SAS system for factor analysis and structural equation modeling.* Cary, NC: SAS Institute.

Hausman, J. A. (1978). Specification tests in econometrics. *Econometrica, 46*(6), 1251–1271.

Hausman, J. A., Hall, B. H., & Griliches, Z. (1984). Econometric models for count data with an application to the patents-R & D relationship. *Econometrica, 52,* 909–938.

Hausman, J. A., & Taylor, W. E. (1981). Panel data and unobservable individual effects. *Econometrica, 49,* 1377–1398.

Honoré, B. E. (1993). Orthogonality conditions for tobit models with fixed effects and lagged dependent variables. *Journal of Econometrics, 59,* 35–61.

Honoré, B. E., & Kyriazidou, E. (2000). Panel data discrete choice models with lagged dependent variables. *Econometrica, 68,* 839–874.

Hsiao, C. (1986). *Analysis of panel data.* Cambridge, UK: Cambridge University Press.

Judge, G., Hill, C., Griffiths, W., & Lee, T. (1985). *The theory and practice of econometrics.* New York: Wiley.

Kalbfleisch, J. D., & Sprott, D. A. (1970). Applications of likelihood methods to models involving large numbers of parameters (with discussion). *Journal of the Royal Statistical Society, Series B, 32,* 175–208.

Kenward, M. G., & Jones, B. (1991). The analysis of categorical data from cross-over trials using a latent variable model. *Statistics in Medicine, 10,* 1607–1619.

Kline, R. B. (2004). *Principles and practice of structural equation modeling* (2nd ed.). New York: Guilford Press.

Kreft, I. G. G., & De Leeuw, J. (1995). The effect of different forms of centering in hierarchical linear models. *Multivariate Behavioral Research, 30,* 1–21.

LaMotte, L. R. (1983). Fixed-, random-, and mixed-effects models. In S. Kotz, N. L. Johnson, & C. B. Read (Eds.), *Encyclopedia of statistical sciences* (pp. 137–141). New York: Wiley.

Lancaster, T. (2000). The incidental parameter problem since 1948. *Journal of Econometrics, 95,* 391–413.

Lewis-Beck, M. S. (1995). *Data analysis: An introduction.* Thousand Oaks, CA: Sage.

Long, J. S. (1983). *Covariance structure models: An introduction to LISREL.* Beverly Hills, CA: Sage.

Long, J. S. (1997). *Regression models for categorical and limited dependent variables.* Thousand Oaks, CA: Sage.

Maclure, M. (1991). The case-crossover design: A method for studying transient effects on the risk of acute events. *American Journal of Epidemiology, 133,* 144–153.

Mooney, C. Z., & Duval, R. D. (1993). *Bootstrapping: A non-parametric approach to statistical inference.* Newbury Park, CA: Sage.

Mundlak, Y. (1978). On the pooling of time series and cross sectional data. *Econometrica, 56,* 69–86.

Muthén, B. (1994). Multilevel covariance structure analysis. *Sociological Methods & Research, 22,* 376–398.

Muthén, B., & Curran, P. (1997). General longitudinal modeling of individual differences in experimental designs: A latent variable framework for analysis and power estimation. *Psychological Methods, 2,* 371–402.

Neuhaus, J. M., & Kalbfleisch, J. D. (1998). Between- and within-cluster covariate effects in the analysis of clustered data. *Biometrics, 54*(2), 638–645.

Pampel, F. C. (2000). *Logistic regression: A primer.* Thousand Oaks, CA: Sage.

Senn, S. (1993). *Cross-over trials in clinical research.* New York: Wiley.

Singer, J. B., & Willett, J. D. (2003). *Applied longitudinal data analysis: Modeling change and event occurrence.* New York: Oxford University Press.

Sobel, M. E. (1995). Causal inference in the social and behavioral sciences. In G. Arminger, C. C. Clogg, & M. E. Sobel (Eds.), *Handbook of statistical modeling for the social and behavioral sciences* (pp. 1–38). New York: Plenum Press.

Suissa, S. (1995). The case-time-control design. *Epidemiology, 6,* 248–253.

Teachman, J., Duncan, G., Yeung, J., & Levy, D. (2001). Covariance structure models for fixed and random effects. *Sociological Methods and Research, 30,* 271–288.

Therneau, T. M., & Grambsch, P. (2000). *Modeling survival data: Extending the Cox model.* New York: Springer-Verlag.

Tjur, T. A. (1982). Connection between Rasch item analysis model and a multiplicative Poisson model. *Scandinavian Journal of Statistics, 9,* 23–30.

Wooldridge, J. M. (2002). *Econometric analysis of cross section and panel data.* Cambridge: MIT Press.

AUTHOR INDEX

Abrevaya, J., 32
Albert, A., 81
Allison, P. D., 5, 43, 44, 48n, 62, 64,
    72, 74, 78, 81, 84, 86n, 91, 95, 96
Anderson, J. A., 81
Arellano, M., 95

Baltagi, B. H., 95
Begg, C. B., 44
Bollen, K. A., 91
Bond, S., 95
Bryk, A. S., 25

Cameron, A. C., 49, 50, 54, 59, 61
Chamberlain, G. A., 32, 44, 78
Christakis, N., 84, 86n
Conaway, M. R., 44
Cox, D. R., 71
Curran, P., 90

Darroch, J. N., 44
De Leeuw, J., 25
Duncan, G., 91
Dunteman, G. H., 5
Duval, R. D., 52

Ejrnaes, M., 98n
England, P., 95

Grambsch, P., 73
Gray, R., 44
Greene, W. H., 14, 64, 86n
Greenland, S., 82, 83
Griffiths, W., 18
Griliches, Z., 49, 62

Hall, B. H., 49, 62
Hatcher, L., 90
Hausman, J. A., 23,
    27n, 49, 62, 98n
Hill, C., 18
Ho, M. R., 5
Holm, A., 98n
Honoré, B. E., 95
Hsiao, C., 32

Jones, B., 44
Judge, G., 18

Kalbfleisch, J. D., 32, 39
Kenward, M. G., 44
Kline, R. B., 88, 90
Kreft, I. G. G., 25
Kyriazidou, E., 95

LaMotte, L. R., 2
Lancaster, T., 32, 95
Lee, T., 18
Levy, D., 91
Lewis-Beck, M. S., 4
Long, J. S., 5, 49

Maclure, M., 79
McCloud, P. I., 44
Mooney, C. Z., 52
Mundlak, Y., 23, 27n, 91
Muthén, B., 88, 90

Neuhaus, J. M., 39

Pampel, F. C., 5

Raudenbusch, S. W., 25

Senn, S., 4
Singer, J. B., 25
Sobel, M. E., 2
Sprott, D. A., 32
Suissa, S., 82

Taylor, W. E., 98n
Teachman, J., 91

Therneau, T. M., 73
Tjur, T. A., 44
Trivedi, P. K., 49, 50, 54, 59, 61

Waterman, R., 62, 64
Willett, J. D., 25
Wooldridge, J. M., 3, 94, 96
Wu, Y., 95

Yeung, J., 91

SUBJECT INDEX

Antisocial behavior example, 8–25, 42–47, 89–93
Average treatment effect, 1

Between R^2, 19
Bias, 4
  in Cox regression with fixed effects, 74, 78
  in fixed effects as latent variable model, 92
  in linear fixed effects models, 23
Binary logistic regression, 42–45
Binomial models, negative, for count data, 61–64, 65 (table), 68n
Birth intervals example, 70–78
Bootstrap standard errors, 51–54, 57, 60, 61
British Household Panel Study Survey, viii

Case-crossover method, 79–82, 84
Case-time-control method, 82–85
Categorical response variables:
  ordered, 42–44
  unordered, 44–47
Censored cases, 83
Censored intervals, 71
Chi-square statistics
  in logistic fixed effects models, 41, 44, 48n
  in negative binomial models, 63, 68
  in structural equation models, 92, 93, 98n
Christakis, Nicholas, 86n

Coefficients
  in linear fixed effects models, 24
  vectors of, 6
Commands in Stata. *See* Stata software
Complete separation, 81
Conditional likelihood, 33, 35 (table), 38 (table)
Conditional maximum likelihood, 29, 32, 80
  in Poisson models for count data, 50, 53 (table), 54, 60 (table), 62, 64
  See also Maximum likelihood estimation
Conditional method, 18
Confidence intervals, 42, 64
Constant variance assumption, 6
Convergence failure, 80–81
Count data, fixed effects models for, 49–69
  hybrid model, 65–68
  negative binomial models, 61–64, 65 (table)
  Poisson models with more than two periods, 54–61
  Poisson models with two periods, 49–54
Country-year example, viii
Cox, David, 71
Cox regression
  with fixed effects, 73–77, 75 (table)
  hybrid methods for, 79
  review of method, 71–73, 72 (table)

Cross-classification, 29 (table)
Cross-lagged coefficients, 96–97
Cross-lagged panel model, 95
Cumulative logit model, 42–44
Current Population Survey: Annual
    Demographic File (CPS), 95

Data
    balanced, 89
    of count variables. *See* Count data,
        fixed effects models for
    of event history. *See* Events history
        data, fixed effects models for
        requirements for fixed effects
        methods, 2
Death of spouse example, 79–85
Degrees of freedom, 92, 93
Dependent variables, 2–4
Deviance statistic, 57, 64
Deviation coefficients, 25, 40–41,
    41 (table), 44, 45, 67
Difference scores, 7–12, 29, 30 (table)
Dummy variable method for two or
    more periods per individual,
    14–17, 15 (table), 16 (table)
    compared to other methods, 17
Dummy variables
    in Cox regression with fixed
        effects, 73
    in linear fixed effects models,
        12, 14
    in logistic fixed effects models, 32
    in Poisson models, 59
Duration analysis, 70
Dynamic models, 95

Education and wages example, 3
Efficiency, 4, 12, 23
Endogenous variables, 88, 94
Error terms, 6, 96
Event history analysis, 70
Event history data, 70
Events, 70
    independence of, 50

Events history data, fixed effects
    models for, 70–86
    case-crossover study, 79–82
    case-time-control method, 82–85
    Cox regression review, 71–73
    Cox regression with fixed effects,
        73–79
    hybrid method for
        Cox regression, 79
Exogenous variables, 6, 88, 96

Failure time analysis, 70
First difference equation, 7, 10
First difference method
    compared to other methods, 17
    in lagged predictors model, 94
    for three or more periods, 12–14,
        13 (table)
    for two-period case, 7–12
Fixed effects regression models
    basics of, 1–5
    compared to random effects models,
        viii, 21–23, 91
    compromise with random effects,
        92–93
    for count data. *See* Count data,
        fixed effects models for
    for events history data. *See* Events
        history data, fixed effects
        models for
    as latent variable model, 91–92
    linear models. *See* Linear fixed
        effects regression models
    logistic models. *See* Logistic fixed
        effects regression models
    for structural equation models. *See*
        Structural equation models
        with fixed effects
Frailty term, 76

Generalized estimating equations
    (GEE) method, 33, 36, 46 (table)
    compared to Poisson estimates,
        57–59, 58 (table)

in count data hybrid method,
66–67 (table), 66–68
Generalized least squares (GLS)
regression, 14, 21, 22 (table)
Generalized logit model, 44–47
Gompertz model, 77, 77 (table), 107
Group mean centering, 25

Hausman test, 23, 27n, 92, 98n, 100
Hazard, 71
Hazard analysis, 70
Hazard ratios, 72–73, 75
Heart attack and statin drug
example, 36
Hybrid methods
compared to structural equation
models, 92
for Cox regression, 79
of cumulative logit model, 42, 43
(table)
of fixed effects models for count
data, 65–68, 66–67 (table)
of linear fixed effects regression
models, 23–25, 24 (table)
of logistic fixed effects regression
models, 39–42, 40 (table)
of multinomial logit model,
45, 46 (table)

Incidental parameters problem,
32, 64, 74
Independence of events assumption,
50, 57
Instrumental variables (IV)
framework, 95
Intensity function, 71
Interactions with time. See Time
interactions
Intercepts, 89
Intervals
birth intervals example, 70–78
censored, 71
confidence, 42, 64

Jackknife standard errors, 51–52, 61

Lagged dependent variable, 93–97
Lagged predictors, reciprocal effects
with, 93–97
Latent variable model
fixed effects as, 91–92
random effects as, 87–90
Latent variables, 88, 96
Likelihood ratio test, 92
Linear fixed effects
regression models, 6–27
compared to other methods,
31, 32, 40, 43
dummy variable method, 14–17
hybrid method, 23–25
mean deviation method, 17–19
random effects models
comparison, 21–23
three or more periods, first-
difference method, 12–14
time interactions, 19–21
two-period case, 7–10
two-period case, difference score
method extended, 10–12
Linear structural equation model, 87
Logistic fixed effects regression
models, 28–48
compared to other methods,
31, 32, 40, 43
cumulative logit model, 42–44
hybrid method, 39–42
multinomial logit model, 44–47
Poisson model converted into, 51
three or more periods, 32–37
time interactions, 37–39
two-period case, 28–31
Log-linear model, 44
Log-odds of poverty, 31

Marriage and recidivism example, 1, 4
Maximum likelihood estimation, 29,
90, 95
Mean coefficients, 25, 41,
41 (table), 44, 67
Mean deviation method, 17–19
Mean deviation variables, 23–25

Monte Carlo simulations, 64
Mortality example, 79–85
Mplus software, 5, 88–90, 93,
    108–112
Multinomial logit model,
    44–47, 46 (table)

National Longitudinal Study of
    Adolescent Health, viii
National Longitudinal Survey of
    Youth (NLSY)
    in linear fixed effects models,
        7, 10, 12, 19, 24
    in logistic fixed effects models, 28
    in structural equation models,
        89, 90 (table)
National Survey of Family Growth
    (NSFG), 70
NB2 model, 61–62
Negative binomial models for count
    data, 61–64, 63 (table),
    65 (table), 68n
Nonconvergence, 80–81
Nonrepeated events, fixed effects
    event history methods for, 79–85
Null hypothesis, 18, 44, 63
    Hausman test of, 23, 27n

Observation data sets, 15 (table),
    34 (table), 55–56 (table)
Observed variables, 3
Occupations example, 95–96
Odds ratios, 81 (table), 81–82, 84 (table)
Ordered logit model, 42–44
Ordinary least squares (OLS)
    regression
    with count data, 49
    in lagged predictors model, 94
    in linear fixed effects models, 7, 8,
        9 (table), 11 (table), 17, 21
Outer product of gradient (OPG)
    standard error, 63 (table), 64
Overall R^2, 19
Overdispersion, 50, 51,
    52, 57, 61, 62

Panel models, 95
Panel surveys, viii
Partial likelihood method, 71–73
Patents from firms example, 49–67
Path diagrams, 88, 88 (figure),
    91, 91 (figure)
Period, 6
    See also Three or more periods per
        individual; Two-period cases
Poisson models for count data:
    with more than two periods per
        individual, 54–60, 55–56
        (table), 58 (table), 60 (table)
        61 (table)
    with two periods per individual,
        49–54, 53 (table)
Population-averaged coefficients,
    36–37, 45, 46 (table), 47
Population-averaged model
    compared to Poisson estimates,
        57–59
    in count data hybrid method,
        66–67 (table), 66–68
Poverty of teens example, 28–41
Predictor variables, 2–4
Proportional hazards model, 71–73

R^2, within, between, overall, 19
Random assignment of variables, 1
Random effects models, 2–4
    compared to fixed effects models,
        viii, 21–23, 91
    compared to Poisson estimates,
        57, 58 (table)
    compromise with fixed effects,
        92–93, 93 (table)
    in count data hybrid method, 66–68
    Cox regression, 76
    Gompertz model, 77,
        77 (table), 107
    as latent variable model, 87–90
    maximum likelihood estimation,
        33, 35 (table), 36
Random intercept models, 25
Random slope models, 25

Recidivism and marriage example,
1, 4
Reciprocal effects with lagged
predictors, 93–97, 96 (table)
Response variables
in logistic fixed effects models, 28
ordered categorical, 42–44
in structural equation models, 87
unordered categorical, 44–47
Robust standard errors, 73

SAS software, 5
ABSORB statement, 18
MIXED procedure, 21, 25
Saturated model, 86n
School performance and video play
time example, 2
SEM. *See* Structural equation models
with fixed effects
Shared frailty models, 76
Software
for fixed effects methods, 4
for structural equation models, 88
See also Mplus software; SAS
software; Stata software
commands
Standard deviation, 42
Standard errors
in count data models, 51–54,
57, 59, 60, 61, 64
in Cox regression, 73
in fixed effects as latent variable
model, 91–92
in linear fixed effects models,
12, 17, 18, 22, 24
in logistic fixed effects models, 36
in structural equation models, 93
Stata programs, 5, 99–107
Stata software commands
areg with absorb(id) option, 27n
blogit, 51, 103–104
clogit, 32, 80
gllamm, 97n
logit, 29, 101, 107
mlogit, 45, 103

nbreg, 62, 69n, 106
nbreg with dispersion(constant)
option, 69n
nbreg with vce(opg)
option, 64, 106
ologit, 43, 102
reg, 16, 27n, 99, 100
reshape, 69n, 99–103, 105
stcox, 72, 106
stcox with shared(caseid)
option, 76
stcox with strata(caseid)
option, 74, 107
stcox with vce(cluster caseid)
option, 73, 107
streg, 77, 107
test, 44, 48n, 76, 101,
102, 103, 106
xtabond, 95
xtlogit, 32–33, 40, 47, 48n, 80,
101–103, 107
xtmelogit, 41, 102
xtmixed, 25, 101
xtnbreg, 62, 64, 66, 106
xtpoisson, 54, 57, 59, 105
xtreg, 18, 21, 23, 27n, 89,
90, 92, 100, 101
xtreg with (pa) option, 14, 99
Stratification, 74
Strictly exogenous variable, 6
Structural equation model (SEM),
88, 90 (table)
Structural equation models with fixed
effects, 87–98
compromise between fixed effects
and random effects, 92–93
fixed effects as latent variable
model, 91–92
random effects as latent variable
model, 87–90
reciprocal effects with lagged
predictors, 93–97
Subject-specific coefficient,
36–37, 46 (table), 47
Survival analysis, 70

Teen poverty example, 28–41, 44–47
Three or more periods per individual
  in linear fixed effects models,
    12–14
  in logistic fixed effects models,
    32–37
  in Poisson models for count data,
    54–61
Tice, Peter, 27n
Time control in case-time-control
  method, 82–85
Time interactions
  in linear fixed effects method,
    19–21, 20 (table)
  in logistic fixed effects models,
    37–39
Time-invariant variables
  in Cox regression, 76
  in hybrid method of fixed effects
    models for count data, 67
  in linear fixed effects models, 8–9,
    10, 15, 19, 21–25
  in logistic fixed effects models,
    31, 37–39
  in Poisson models, 49, 59–61
Time variable, 15
Transition analysis, 70
Two or more periods per individual in
  dummy variable method, 14–17
Two-period cases
  comparisons of methods, 31
  of linear fixed effects models, 7–10
  of linear fixed effects models,
    extending difference scores,
    10–12
  of logistic fixed effects models,
    28–31

in Poisson models for count data,
  49–54
Two-wave, two-variable panel
  model, 95, 97

Unconditional maximum likelihood, 32
  for negative binomial models,
    62–64, 63 (table)
  for Poisson models for count
    data, 54, 59
  See also Maximum likelihood
    estimation
Underidentified model, 91
Unobserved variables, 3
Unordered categorical variables, 44–47

Variables in fixed effects models,
  1–4, 6
  endogenous, 88, 94
  exogenous, 88, 96
  mean deviation, 23–25
  See also Categorical response
    variables; Dummy variables;
    Response variables; Time-
    invariant variables
Variance, proportion of, 18
Vectors of coefficients, 6
Video play time and school
  performance example, 2

Wages and education example, 3
Wald chi-square, 68
Wald test, 25
Within R^2, 19

Zero-inflated Poisson models, 68n
Zero mean, 6

Supporting researchers for more than 40 years

Research methods have always been at the core of SAGE's publishing program. Founder Sara Miller McCune published SAGE's first methods book, *Public Policy Evaluation*, in 1970. Soon after, she launched the *Quantitative Applications in the Social Sciences* series—affectionately known as the "little green books."

Always at the forefront of developing and supporting new approaches in methods, SAGE published early groundbreaking texts and journals in the fields of qualitative methods and evaluation.

Today, more than 40 years and two million little green books later, SAGE continues to push the boundaries with a growing list of more than 1,200 research methods books, journals, and reference works across the social, behavioral, and health sciences. Its imprints—Pine Forge Press, home of innovative textbooks in sociology, and Corwin, publisher of PreK–12 resources for teachers and administrators—broaden SAGE's range of offerings in methods. SAGE further extended its impact in 2008 when it acquired CQ Press and its best-selling and highly respected political science research methods list.

From qualitative, quantitative, and mixed methods to evaluation, SAGE is the essential resource for academics and practitioners looking for the latest methods by leading scholars.

For more information, visit **www.sagepub.com**.